W9-BGL-969

Soil: The Skin of the Planet Earth

Miroslav Kutílek • Donald R. Nielsen

Soil: The Skin of the Planet Earth

 Springer

Miroslav Kutílek
Czech Technical University
Prague, Czech Republic

Donald R. Nielsen
Land, Air and Water Resources
University of California
Davis, CA, USA

ISBN 978-94-017-9788-7   ISBN 978-94-017-9789-4   (eBook)
DOI 10.1007/978-94-017-9789-4

Library of Congress Control Number: 2015935433

Springer Dordrecht Heidelberg New York London
© Springer Science+Business Media Dordrecht 2015
This work is subject to copyright. All rights are reserved by the Publisher, whether the whole or part of the material is concerned, specifically the rights of translation, reprinting, reuse of illustrations, recitation, broadcasting, reproduction on microfilms or in any other physical way, and transmission or information storage and retrieval, electronic adaptation, computer software, or by similar or dissimilar methodology now known or hereafter developed.
The use of general descriptive names, registered names, trademarks, service marks, etc. in this publication does not imply, even in the absence of a specific statement, that such names are exempt from the relevant protective laws and regulations and therefore free for general use.
The publisher, the authors and the editors are safe to assume that the advice and information in this book are believed to be true and accurate at the date of publication. Neither the publisher nor the authors or the editors give a warranty, express or implied, with respect to the material contained herein or for any errors or omissions that may have been made.

Cover illustration: © stocksolutions/foltolia.com

Printed on acid-free paper

Springer Science+Business Media B.V. Dordrecht is part of Springer Science+Business Media (www.springer.com)

Contents

Chapter 1
Omnipresent Soils

Knowledge of our *Homo sapiens* existence is restricted to a tiny patch of both space and time in the cosmic infinity. In spite of the presence of such infinite dimensions and our feelings of being isolated, lost, and completely out of touch with all other vast, unknown reaches within and outside the universe, it will help us to put parts of our known world into mutual relationships. Our soils form a very thin peel on the surface of our planet Earth. The total surface area of Earth is 510 million km^2 or 5.1×10^8 km^2. Real soils have been born on terrestrial landscapes that extend across 143,330,000 km^2 or rounded to 1.43×10^8 km^2 when we omit the area of Antarctica covered by ice during the last 2 million years. The surface area of all planets in our solar system, rounded to 1.22×10^{11} km^2, is roughly 3 orders of magnitude larger than the area of all our continents and islands, again without Antarctica. Or, in other words, the soil covering our Earth's solid surface is a thousand times smaller than the surface areas of all planets in our solar system.

The size of the surface area of all soils on our planet Earth is not a single fascinating property. There are countless processes running in that thin peel that they remind us of numerous functions running in the skin of living bodies. The comparison of soil to the skin of animals was for the first time used in 2005 by Alfred Hartemink, Secretary General of the IUSS, in the brochure prepared for the occasion of the International Year of Planet Earth. The booklet's title was *Soil – Earth's living skin*. It is stated there that without soil *the Earth's landscape would be as barren as Mars*. Awed by the parable *soil – skin*, we modified it in our book's title.

Soils evolved on Earth's continents having two specific properties, both of which are difficult to grasp and fully appreciate in terms of our everyday and lifetime experiences. One is the extremely large surface area of solid particles within the soil, e.g., if we measure it in the top 30 cm of soil below a land surface area of 1 m^2, we obtain an average value of about 10^6–10^7 m^2. It is the size of about a square kilometer or even more. The second soil property is the magnitude and importance of the hollows between the solid particles called soil pores. These pores usually occupy about 50 ± 10 % of the total soil volume and have sizes ranging between tens of nanometers up to hundreds of micrometers. The volume of millimeter-sized pores is

© Springer Science+Business Media Dordrecht 2015
M. Kutílek, D.R. Nielsen, *Soil: The Skin of the Planet Earth*,
DOI 10.1007/978-94-017-9789-4_1

usually negligibly small. The total pore space is full of life and chemical reactions and provides a variety of pathways for storage and transport within and through each soil. Muddy water is "purified" during its flow through soil. Nutrients important for the existence of plants are dissolved in soil water and enter together with water into roots. The soil pores are spaces where microbes, microscopic fungi, and all forms of life simultaneously perform various reactions and transformations inside of soil and especially in the vicinity of plants' roots. The global distribution of all soils is the essential key for water to circulate in all directions of the Earth's water cycle that we usually take for granted.

The soil could emphatically declare, "I am the protective filter and the mediator of energy, I am responsible for transformation of inorganic and organic compounds, I am the sustainer of productive life, and I am the cradle of man's life and culture. At the same time I am the medium for deposing the dust remaining after the death of man. I am the myth."

The soil could be proud that its qualities were most advanced in Egyptian mythology. The Egyptians discovered divinity in every part of environment and especially in happenings for which they had no explanation except of mythology. Just from the start of permanent settlement of the Nile valley, their experience did not predict the height that a Nile flood would reach and therefore what harvest they could expect. Low discharge meant a catastrophe especially when it happened in a series of several years and soils did not receive the needed portions of water and of fine fertile mud. But extremely high discharge resulted in catastrophe, too, with all signs of catastrophic floods. The combination of Nile discharge observation with some social acts or without current behavior of certain animals led to the birth of magic. Finally, the ritual of soil fertility amelioration was believed to be reached by repeating unusual acts or by killing animals as the bearers of oddities. The key factor of myths was the substitution of chaos by order, replacement of chaos by balance of natural forces, which were not described in an abstract form but as gods.

Ra was the great sun god at Heliopolis (Lunu in old Egyptian or On in Coptic). Shu and Tefnut were his children and Geb and Nut were their offspring. Geb was the god of earth and of soil. Osiris, the first child of Geb and Nut, was a god of nature and vegetation, and thus, he was also close to soils. He represents the Nile with its annual flooding and withdrawal; his sister goddess Isis represents the fertile farmland of Egypt, which was made productive by the Nile; one of his brothers, Set who personified evil, represents the arid desert that is separated from the Nile and the fertile land along the Nile, while his sister, goddess Nephthys, represents the marginal areas between the farmland and desert. This separation of soils was actually the first classification of soils.

Classical Greek myths speak about Demeter, the daughter of Kronos who swallowed her together with four other sisters and brothers. When she was liberated by Zeus, she was asked to take care of grain, farming, and soils. But her part among the gods started to be much more complicated. Her daughter Persephone was abducted by Hades, the son of Zeus and god of the underworld. Because she could not escape back from the underworld, her mother Demeter being extremely sad withdrew to her temple and changed the earlier fertile soils into complete infertility. The starving people stopped sacrificing to gods. With a catastrophe being imminent for mankind as well as for the world of gods, Zeus had to propose a compromise to Demeter. Persephone will spend two thirds of the year with her mother in the world and one third of the year with Hades in the underworld. Her myth explains the seasons: plants grow and bear fruit while Persephone is aboveground with her mother, but wither and die during the months she spends with Hades. Demeter instructed the first man (or Eleusinian demigod) Triptolemos how to take care of soil – how to plow, sow, and harvest. Since Triptolemos and his brother Demophon taught all people the magic of agriculture and how to keep soils fertile, they were worldwide famous by their knowledge.

In the Old Testament we find many sentences dealing with soils and their fertility. Yahweh said to his people, "Take care that the land be able to support you, when your days and your children's days are multiplied" (Deuteronomy 11: 16–21). Yahweh reminded to man that he should till it and keep it and when working he will be sweating. Hard toiling will be associated with man's cultivation of the soil (Genesis 3: 17–19). In Genesis 3: 17–19 God said to Adam: "All your life you will sweat to produce food, until your dying day. Then you will return to the ground from which you came. For you were made from dust, and to dust you will return." It shows the intimate relationship of humans and soil. In some other passages soil and water stand side by side in their importance. For example, the phrase "… like a well-watered garden…" represents fruitfulness (Isaiah 58: 11).

We find concrete instructions, e.g., in Exodus 23: 10–11a: "Plant and harvest your crops for six years, but let the land rest and lie fallow during the seventh year," or in reports about the past: Genesis 47: 23–24 reports that Joseph supplied seed for the farmers at the end of the great drought so that they could continue to farm as good years returned, or in Ezekiel 17: 5: "Then he took some of the seed of the land and planted it in good ground for growing."

Without studying historic scripts, we understand the importance of soils not only for our recent civilizations but also for all forms of life on our planet.

The soil surrounds plant roots and their growth is decisive for plants and their healthy growth is, in turn, also decisive for our life. Joan Miró, the famous Catalan Spanish surrealist and dadaist painter and sculptor, said that if we like to be real artists, we have to put down roots, which meant that an artist has to find his own style and when he succeeds, he must keep, extend, and explore it like the plant which is faithful to the fitting soil, or vice versa the soil offers the best support to the fitting plant.

Without exaggerating, we soil scientists could say that there are as many life forms in a handful of soil as there are humans on Earth. And everything – from catching the first raindrops to food production – depends on the soil. If we have studied the appropriate facts and if we now succeed in explaining them, we ought to do what is appropriate and correct with that knowledge. The jump from knowledge to actions, however, is often a huge leap – but without knowledge, it is impossible.

The study of the soil has its own scientific name *pedology* that was derived from the classical Greek words *pedon* meaning ground and *logos* meaning reason or knowledge. Its derivation had nothing to do with Latin *pes, pedis* (genitive case), and *pedes* (plural) meaning foot, with Greek *pais* or its shortened form *ped* meaning child, and with Greek *ped* meaning education in encyclopedia and in pedagogue meaning a child's guide. Persons studying soils are called pedologists and should not be confused with those known as pediatricians or pedophilia.

The term pedology was probably used for the first time and only once and formally in the title of the book *Pedologie oder allgemeine und besondere Bodenkunde* by F.A. Fallou and G. Wiedemann in 1862. However, the scientific approach to the principles of origin and properties of soils was still missing. We had to wait for 20 years more until the Russian V. Dokuchaev formulated the scientific principles of origin and properties of soils. His role in foundation of soil science, or pedology, has been appropriately described by A. Hartemink, who compared Dokuchaev's unique creativity to making the first Cremona violin – already perfect and ready to use since his fieldwork was accompanied by the theory and vice versa. As a result, we are all disciples of Dokuchaev.

Most people living in cities and urban communities who spend a great majority of their time in offices of institutions and in factories are used to saying "Soil is dirt." Those words are virtually never spoken by farmers nor heard from the mouths of persons living in rural areas who appreciate the local and regional essential importance of soils. We soil scientists are aware of the universal importance of soils and readily understand that without soil, life in its contemporary form would not exist today. We are organized in local clubs and national societies as well as in the International Union of Soil Science. Our cooperation in research and exchange of new theories as well as discussing the application of theories in farmers' field, the amelioration of soil properties for practical aims in water management, and the preservation of healthy soils and maintenance of profitable environments are realized at conferences and courses. The publication of papers in international journals is an inseparable part of our specialized studies. Here, we list the names of only a few of the many frequently cited journals: *Soil Science, La ciencia del suelo y nutrición vegetal, Journal of Plant Nutrition and Soil Science* (*Zeitschrift für Pflanzenernährung und Bodenkunde*), *European Journal of Soil Science, Catena, Geoderma, Soil Science Society of America Journal, Soil Technology* that was incorporated in *Soil and Tillage Research*, and many national journals of soil science, like *Canadian Journal of Soil Science* or *Japanese Journal of Soil Science and Plant Nutrition* (*Nippon dojohiryogaku zasshi*).

If we could understand the soil language, let us say Pedolang or Pedenglish, we would hear: "Hi, all of you *Homo sapiens*. Do you know me? I am the membrane on the top of Earth. I am thinner than apple peel and I have more functions than that peel. I am protecting your life, I am filtering the dissolved compounds and the finest particles from water and thus I am assuring your safe life. I am your buffer smoothing energy oscillations and balancing water excess with water shortage during droughts. I am your sustainer of life, your supporter. I have endless capacity to store useful compounds and elements and to offer pleasant environments for the genesis of new forms of life. You *Homo sapiens* should be aware of what I have just said in

Pedenglish. Therefore, today I am the soil who is supporting two soil scientists in their effort to offer you – the reader of this book – all the available information about me, the **soil**."

And we, Mirek Kutílek and Don Nielsen, shall try not to frustrate you, our soil, who has been with us continuously since we first opened our eyes.

Chapter 2
How I Earned My First Crown and Dollar

2.1 The Story of the Author Kutílek

While attending high school, I had to write an essay on fate. Now, after more than half a century, my recollection is for sure not exact, but here I at least try to keep the atmosphere:

The famous economist Antonio Usurero, who missed the Nobel Prize by just the thickness of a hair, wrote in one of his books: "…. The most important affair in our life is from how we earned the first dollar in our childhood or as teenagers. This is a sort of prediction by fairies on our future fate and destiny. It is lasting forever and it is unavoidable." Since we have been taught that the best brains of mankind are to be trusted, I am going to find out what will be my future profession.

I spent my best childhood vacations in a small Czech village *Hlinoviste* and the rough English translation will be either "Loam Site" or "Loam Pit" since loam is *hlina* in Czech. My grandma was the co-owner of a pub and of a small farm there together with my uncle, who was single at that time. During harvest my grandma and uncle were obliged to be in the field, but they had to keep the pub open for the old men of the village who were no longer farming, but nevertheless typical of how all farmers used to be – thirsty not only in the evenings but throughout the entire day. After I was given the keys of the pub and its cellar, my duty was to tap beer and cash the guests if they were leaving. Being a student in fourth grade and excellent at adding, subtracting, and simple arithmetic, I had no problems cashing the guests. If they paid by notes, it was an easy task for me to give their change back in smaller notes and coins.

When my grandma came back from the field, she found some of old guests still in the pub.

With his index finger pointing at me, one of the group of old men at the table reserved for regular guests said, "You have a perfect headwaiter!"

"Do you think so?," asked my grandma with pride being felt in her voice.

"How old is he?," asked the curious old man.

© Springer Science+Business Media Dordrecht 2015
M. Kutílek, D.R. Nielsen, *Soil: The Skin of the Planet Earth*,
DOI 10.1007/978-94-017-9789-4_2

"About five," answered my grandma, whose memory was chaotic when discussions were about age. Actually, she stopped counting my years when I was 5 years old.

"He is a genius," exclaimed the thin old man, and all guests repeated in chorus: "Genius!" From this moment on I was the little genius for all guests from our village Loam Pit. When their bill was about three crowns and several hellers, they paid five crowns and they left me more than one crown tip and I was the best tipped waiter in the county side.

My fate is to be the headwaiter, I wrote at the end of the essay.

Our high school teacher was beautiful with a perfect makeup. All of my friends fell in love with her and the more we recognized our hopelessness, the cheekier we were. She asked me: "Where did you read about the economist Usurero? Never heard about him."

"He is well known among professionals in theory of economy," I was lying. I would never admit that I derived this imaginary name from the Middle English word *usurer* that means a lender of money at huge, unlawfully high interest rates. Knowing that the Latin *usura* means loan, I invented the Italian family name of Usurero by adding an "o" to the end of usurer. I also remembered from history that after Italians started banking, the somewhat dirty term bankruptcy stemming from Italian *banco rotto* meaning broken bench became a common word. In those times, bankers illustrated their ability and readiness for making loans by sitting on a box-banco containing available cash. They rotated the big box up and down to demonstrate the sad situation whenever cash was no longer available for more loans. While I was writing my essay at the high school, I decided to improvise the name of the economist as well as his well-known professional recognition.

I was pleased when she responded, "Strange name. But you wrote the essay well."

After that, everybody at school started to call me *pingl*, a Czech vulgar expression for a trainee waiter.

It took me more than 10 years to recognize that my nonrealistic quotation of an imaginary Usurero was correct but in a different way than deriving my fate from my pub expertise. My imaginary Usurero predicted correctly that my lifetime fate would be related to the name of the village where my granny owned the pub. With that name being Loam Pit, my fate was not only linked up with loam but with all soils. Although I started my university study in civil engineering, division of water management, after I recognized what an important role soil plays in the hydrologic circle, I decided to shift my studies into the direction of soil hydrology and soil physics. It happened during the period when many basic equations of water flow in soils were formulated, and on many occasions, I felt like I was living in a scientific thriller as I watched all sorts of mathematical magic being performed by my slightly older colleagues. Because I wished to be a similar magician, soils and water together with their governing physical laws attracted my attention for the rest of my life. This is also why I decided to share my experience with lay readers in telling them about the really magical role of soil in all life forms on our planet Earth. My longtime and best friend Don Nielsen is not only accompanying me, he has often a leading role.

2.2   The Story of the Author Nielsen

During and following the Great Depression (1929–1939), I spent my early youth living in the developing town of Phoenix located within the arid, dusty region of Arizona in Maricopa County across which the dry bed of the Salt River has been existing for centuries. Today, my memoirs actually agree with the statement, "The most important affair in our life is from how we earned the first dollar in our childhood or as teenagers," creatively derived by Kutílek who invented and fictitiously quoted a famous economist named Antonio Usurero.

While behaving and minding my parents and doing a few designated chores in and around our 2-bedroom wooden home, I never thought of working for money because my whims for occasional extra enjoyment were always bought with a few cents given to me by my parents. Some of my friends having rich parents were given weekly allowances to buy ice cream, candy, and junk. But without spending a penny, I enjoyed walking through uninhabited desert regions observing plant and animal life together and also picking up archeological artifacts from Hohokam Indians that I always found on different kinds of soil surfaces. During any of my treks, I often saw javelina, burros, coyotes, wolves, turkeys, buzzards, as well as smaller creatures such as turtles, lizards, rattlesnakes, horned toads, scorpions, tarantulas, ants, spiders, centipedes and millipedes, crickets, earthworms, etc. Having also frequently found archeological artifacts on and below soil surfaces, I saved arrowheads made of flint, stones shaped and used as tomahawks, various grinding stones and bone awls, pieces of turquoise jewelry, ceramic and adobe figurines, and fragments of decorated pottery.

As the depression gradually ended, my desire steadily expanded to enjoy costly activities. Without wanting to further empty the pockets of my parents, I sought any kind of a job to earn some money. Luckily, a part-time job suddenly appeared that I could do each day after school and on weekends. Surprisingly, the surroundings of my first paying job were similar to those encountered on my treks walking on dry soil through the dusty desert environment. Working indoors in a retail store filled with books, baggage, and suitcases that were continually being covered by dust blown into the store from unpaved streets and fallow soils in the sunbaked vicinity, I earned my first dollar repeating what I had already done for years outdoors in the desert. In both cases, I sorted, picked up, individually dusted off, and rearranged each of the objects. In the desert, they were living organisms or artifacts at or near the soil surface. In the store, they were books and baggage that I nicely rearranged after dusting each of them and also after removing lice and silverfish potentially harmful to the books as well as killing unwanted insects, spiders, and rodents.

My first full-time job also involved dusty conditions – digging and sampling soils across farmers' fields to ascertain deficient levels of plant nutrients and also searching for unwanted pests that reduce crop yields. By the time I was about to finish high school, I was happy breathing dusty air while digging in dirt working with farmers. After telling my father, who was a farmer during his entire life, that I had decided to study agriculture in college, he gasped and emphatically said,

"Never! Do you really wish to follow my footsteps working every day from sunrise to long after it's dark without ever having time or making enough money to take a decent vacation? I strongly advise you to study accounting, economics, or some topic to make money. Do not study agriculture and become a farmer like me."

Two months later and convinced that I should strive to become a wealthy accountant or business manager, I entered college. During the first term I took these courses: accounting, economics, business mathematics, sociology, and history. By the middle of the term, I knew that I made a mistake even though I made excellent final grades in all five courses. By the following term, switching gears from money to science, I took botany, chemistry, entomology, geography, and geology. And 3 years later at a different university, I graduated with a BS degree in Agricultural Chemistry and Soils. But with that knowledge and experience gained in classrooms and laboratories and across various landscapes, the exact meaning, behavior, and importance of dust remained somewhat of a mystery to me in relation to the plant and animal life that I had observed in the desert as a youngster. Being curious, I continued my science-related education by exploring the impact of dust and soil particles on microbial communities living within desert topsoils. My exploration was enhanced by using newly available radioactive elements to determine critical levels of carbon, nitrogen, and phosphorus being manipulated by millions of soil microorganisms living in the vicinity of each and every root of a plant. Their dominance controlled the fate of each plant – its metabolism, growth, survival, and reproduction – as well as communities of plant species that thrived or were exterminated on each soil across the desert. Earning an MS degree in soil microbiology was exciting – it opened my eyes and improved my understanding of what I could not see as a teenager without a powerful microscope.

My curiosity continued regarding my early observations of various kinds of animal communities thriving in dust-laden arid regions without any obvious sources of readily available water. With the bulk of each of their individual bodies being composed of water that tends to evaporate daily, where did they find water in desert regions with rain limited to 1 cm per month? Such rainfall seldom provided enough water to accumulate in creek beds that remained dry throughout the year. Not understanding how water infiltrated into and migrated through desert soils nor how communities of micro- and macro-sized animals meandered through and between local hydrological regions of arid to humid environments, I switched my attention to the impact of soils and water on the diversity of animal life by studying soil physics – a combination of soil hydrology, mathematics, and physics conceptually integrated with the sun's energy at the soil surface. Four years later, I earned the PhD in soil physics after analyzing infiltration and redistribution of water within five different field soils using the first homemade portable neutron soil water content measuring device; I continued my childhood habit of walking across the landscape and collecting historical artifacts from the soil. At that time, being nearly 30, married and a father, I was well on my way to fulfill the fictitious quote of Antonio Usurero.

Having lived in only two regions, the desert floor of Arizona and the corn belt of the USA, during the next 20 years I learned more about life on Earth by walking across and digging holes in soils developed under different climates on all continents except Antarctica. Although each trip offered an opportunity to learn something new, every trip ended with my books, baggage, and suitcases needing a thorough dusting just as if I were still earning my first dollar in the bookstore.

Only halfway through my career and still learning, my academic life received a once-in-a-lifetime boost as a result of meeting Kutílek during an international scientific meeting. Although he may assert that when we met he belonged to a country in the underdeveloped part of the world of sciences, he was at that time and remains today a contemporary leader for explaining the evolution of plants and animals including *Homo sapiens* and their adaptability to the ever-changing conditions of soils and global climate. Before meeting him, it never occurred to me to seriously include long-term geologic processes associated with soil genesis, paleopedology, climate change, and archeology that had impacted contemporary soils and their living communities. And as I walked across and dug into dusty soil surfaces around the globe, I never thought of myself as being a member of the living community that I sought to understand.

The second half of my career, filled with many visits to outdoor environments examining soil profiles, fossils from the past, and artifacts stemming from prehistoric communities, was absolutely exquisite owing to my unique inspiration from frequent communications with my greatest personal friend, Kutílek. I even returned to Arizona to walk once again down to the bottom of the Grand Canyon, but at that time, to observe different geologically buried soil profiles, to study remnants of deteriorated Native American villages, and to pay more attention to the impact of the Colorado River eroding and cutting through a region that began to uplift 75 million years ago. And of course, each visit ended with the necessity of removing the dust that accumulated on my baggage and me.

As I recall my lifetime activities, I now believe that the statement attributed to Antonio Usurero by Kutílek was absolutely true – not fictitious. I was born in a dusty environment; earned my first dollar in the middle of a dusty room; spent my entire career studying the intricate complexities, movements, and reactions of dust in living and inert entities on the Earth's surface; and today still learn more about soil without focusing on activities to become rich or to make lots of money. I have always and happily followed an exploratory path directed into soils. I have no intentions to stop in the future until, like other living global organisms, my lifeless residues rejoice within the soil and other domains of the Earth's captivating environment.

Chapter 3
Soil Is the Skin of the Planet Earth

We have mentioned earlier in the introductory chapter that recent soil scientists like to say in their scientific jargon that the soil is a sort of skin to our planet Earth. When we have now more space and the reader has more time, we can afford the luxury of going into some details. First, this comparison is not exact because we mammals are born with skin, have it at the beginning of our existence, and continue to have skin throughout our lives even if it is not in perfect condition upon death. Hearing the parable about Earth's skin, others may deduce that soil and the Earth were created simultaneously. Such a conclusion is not valid. The soil, or the Earth's skin as we used to say, started its existence when macroscopic life was moving from oceans to the mainland, to the surface of continents and islands. This migration was happening roughly 500 million years ago – maybe even a little bit earlier. It was the time of Earth's adolescence and certainly not immediately after the birth of our planet. During this period, various kinds of proto-soils gradually began to uniquely develop and slowly appear at diverse locations. Before this time, only weathering fragments and remnants from rocks – stones, gravels, sands, and even clays – scattered across continental surfaces, completely void of any action by living macroorganisms. Without the contributing actions of these organisms, a soil cannot exist. Although some microorganisms, mainly bacteria, were thriving at that time, their contribution to the transformation of the weathered inorganic rocky material was negligible and has not been documented. As our initially skinless planet aged, an outer jacket of soil eventually became a reality owing to the essential actions of macroscopic life. Real soil does not exist without such a living community.

Another snag or incongruity can be identified in the everyday analogy that compares the relative thickness of the soil on the Earth to that of the skin of humans or animals. The soil layer forming the boundary between the Earth and the atmosphere has a thickness usually of only 1 m, sometimes 2 m, and rarely more than 2 m. This layer in the context of the Earth's radius of 6,378 km is extremely thin – it is about six million times smaller than the radius of the Earth. Sometimes the soil layer is about three million times smaller than the radius of the Earth. Generally, we can say

© Springer Science+Business Media Dordrecht 2015
M. Kutílek, D.R. Nielsen, *Soil: The Skin of the Planet Earth*,
DOI 10.1007/978-94-017-9789-4_3

that the soil's thickness is about million times smaller or six orders of magnitude less (10^{-6}) than the radius of Earth.

On the other hand, the thickness of a human's skin is a thousand times smaller than a human's height, or it is by three orders of magnitude smaller (10^{-3}). The relative thickness of the protective layer of the Earth is a thousand times less than that of humans. In order of magnitude it is 10^{-3}. Comparing the relative thickness of a human's skin to that of the soil, humans are protected much better by their skin than the Earth is protected by its thin soil layer. When we consider a broad number of continental mammals of all sizes, we learn that the relative thickness of their skin is somewhat smaller than the human skin but never falls below a relative magnitude of 100. With the relative soil thickness of Earth being substantially smaller than the relative skin thicknesses of both humans and mammals, we expect that the soil protects the Earth less than the skin protects either humans or mammals. We must take into consideration this expectation of vulnerability when we start discussing the birth, longevity, and death of soils and realize that even a very thin soil is still an unconditional requirement for life on Earth. Regardless of the thickness and continuity of the Earth's topsoil, many geographic locations are not acceptable or conducive for human life owing to local environmental conditions causing extremely thinned or even destroyed skin. We should always remember that the skin of a mammal is protecting just that particular species, while the skin of the Earth is ensuring all forms of life on all continents of our planet.

Our opponents who may not consider the necessity of the myriad of biological processes contributing to the health and safety of the Earth's skin could raise objections against our relative thickness estimates of soils. Merely focusing on physical processes occurring within the transition from the Earth's rocks to the atmosphere above its surface, they may well limit their thoughts to transport between solid and gaseous states in the absence of biological processes. If we accept their statement that the rough average thickness of the Earth's solid crust is about kilometers, we calculate that the actual measured soil thickness of 1 m is more than hundred thousand times smaller than the solid Earth's crust (i.e., in order of magnitude 10^{-5}). Compared to human's skin thickness, it is hundred times less. Considering the mammals' skin, the soil is between ten times and hundred times less than the mammals' skin.

Let us first simply assume that there is a linear indirect relationship between the thickness of the skin and its vulnerability. Then the extremely fine thickness of the Earth's skin – the soil – would mean that the most important Earth's property would be very imperiled if the soil characteristics strongly and abruptly changed. It remains now to show what we consider is the most important property of the Earth. Being egoistic members of a biological order and desiring the sustainability and continued development of positive life conditions for ourselves, we designate the most important property on the prerequisite that it is subject to evolutionary principles. Any abrupt change in the characteristics of the Earth's skin should therefore be avoided. Accepting this principle, soil has a decisive, ever-present influence upon all forms of life activity on Earth. It is a diverse global system of sustainable yet ever-changing local and regional environments that support and delineate micro- and macroorganisms within and across all landscapes. For each location, soil facilitates roots from a plant or a canopy of plants to be anchored in a specific location to thrive and reproduce. Soil together with its vegetative cover keeps the contemporary gas composition of the atmosphere relatively constant – a stable condition required for the existence and survival of human life. Soil retains water from irregular rainfall and irrigation events and offers it with dissolved essential nutrients to plants through their roots in a remarkable, unswerving manner. Soil is the home of countless microorganisms that cause the decomposition and transformation of decayed organic bodies with some of them contributing to the fixation of atmospheric nitrogen. Other microbial species cooperate with roots to such an extent that we could say that they are grazing on the roots causing at the same time a profit for the plant. This type of symbiosis helps both sides – the plant and the microbes. In addition to the activity of microorganisms, we find that soil is the home of ants, termites, earthworms, and huge numbers of related macrofauna that all contribute directly or indirectly to the global vegetative community. Plants could not exist without this continuing chain of influence. The life of herbivores is impossible without plants and the same is valid for us, even if we succeeded in transforming mankind from omnivores to herbivores. And we are not speaking about the majority of us who reject the idea of only veggies.

The modern vocabulary of ecologists permanently retains the term biodiversity. It means the degree of variation of life at all biological levels starting from cellular level up to plants and animals. Biodiversity depends upon the favorable conditions for existence and for evolution of all forms of life. It is our pleasant duty to mention that the biodiversity in soil is much greater inside of the soil than above the soil surface among plants and all animals. Those among the ecologists who are so conscientiously worried about biodiversity should be careful about soil in a same way or even more as they are about living organisms. Why? Because if the biodiversity of soil were lost, an immediate loss of biodiversity on the entire planet Earth would follow. We expect ecologists to preferentially protect soils from their misuse and from the potential, eventual destruction of soils' ability to support life on Earth.

However, many commonly accepted practices of managing soils are destructive even though laymen do not recognize them. A typical example is the planting and cultivation of monocultures that allows a single crop or plant to be grown in a farmer's field or similar areas for a large number of consecutive years. Such a practice steadily leads to the exhaustion of a certain nutrient or group of nutrients which become the most important factor for plant growth with harvests becoming more and more reduced year after year. Even when the availability of the essential nutrient is increased by an annual supply of mineral fertilization, one or more other required nutrients or substances start to diminish or virtually disappear. Eventually with continued monoculture, this lack of irreplaceable matter causes a weakening of the root system, a reduction of growth of the aboveground plant parts, and a deterioration of the effectiveness of the physical, chemical, and biological properties of the soil. However, this result is only one part of the ugly face of monocultures.

Repetitiously cultivating the same crop on the same field also enhances the development of weeds and plant diseases because the natural processes of plant protection are gradually reduced until they approach zero. To combat these unwanted developments and intrusions, farmers apply various kinds and combinations of herbicides and pesticides with some timely short-lasting success whenever they obtain a reasonable harvest. However, even with highest selective agrochemicals yet to be produced, their application in scientifically recommended amounts impacts and even kills countless numbers and kinds of beneficial soil micro- and macrofauna and soil micro- and macroflora – all of which were not the weeds and plant diseases targeted by the farmer. In simple terms, monoculture paves the way for soils to become sick and stripped of their biodiversity.

We will show later on how this plundering of soils leads to the deterioration of all features of the landscape. The negative consequences are not just linked to local catastrophes occurring in the past. We find even today a small-minded approach to grow plants solely for their organic content in order to produce biofuels. They are frequently grown for many years in the same field on an originally very productive soil. Such monoculture fields initially produce abundant yields that are easily harvested and immediately sold for high profits. By continuing such monoculture, we are repeating the same error of our grandfathers, who on so many farms transformed a very fertile soil to a nearly unproductive soil. The recent approach producing a large portion of the biofuels now available is an example of fighting fire with fire. Although we are trying to be less dependent on fossil fuels by producing biofuels, we are forgetting, or perhaps we do not even realize, that we have an undesirable by-product. That by-product is soil having a substantially reduced biodiversity that is not immediately recognized nor considered as important as when a particular species of mammals disappears. The picture and story of the poor animal is published all over the world, while soil plundering is only mentioned in scientific and professional journals. Who among the countless nonprofessionals throughout the world has ever seen any photography of perished soil microbes or fungi? Without speaking about a complete movie, there has never been even a single photo transmitted to any household television set about the disappearance of a microbial colony. It could be that those exact microorganisms in the disappearing colony had been protecting plant roots from attacks by pathogens. Or within the disappearing colony, at least some groups of microorganisms were guaranteeing the beneficial disintegration of decayed plants and animals and the transformation of decomposition products into new organic complexes important for the quality of soils. If those groups disappear, the quality of soil drops sharply. Therefore, it is the biodiversity of soil organisms that is important for soil survival and thus for human civilization.

Christian religions have accentuated the importance of soil by both word and parable. Mankind started with Adam in the Old Testament. It is the word derived probably from the Hebrew *adamah*, a feminine word denoting the earth. Another word *ahava* meaning love is similar to the name *Eva* that means life and it is close to *hayya* that means alive. By combining *Adam* and *Eva* we obtain the living earth full of love. To avoid just a celebration of soil at the start of mankind, we have to mention that *to ʿevah* also means abomination. But it may also mean dirtiness, and if it does, we are back to our soil. Generally, there are several other related meanings coexisting like melodies in the sound from an orchestra.

Soils are not everywhere the same. They change remarkably even along small distances, and when climatic and vegetation conditions change substantially across large distances, their physical, chemical, and biological properties differ to such an extent that their similarity is absent, and soils even display huge assortments of different colors across local, regional, and continental landscapes. Variations of soil are neither chaotic nor by haphazard and correspond to strong relationships imposed by spatial and temporal environmental conditions according to scientific laws. Soils in tundra regions differ substantially from soils of steppes or prairie lands, and they again diverge from soils in humid tropics. Even a nonspecialist recognizes differences according to the color of the soil in a trench or an excavation for a road. A layman notices that tropical soils are usually not gray or grayish brown but much more colorful and sometimes reflect colors of the entire spectrum of a rainbow. Soils differ substantially not only within great distances of climatic and vegetative zones but also within much smaller distances. For example, soils vary along the slope as well as within the valley at the bottom of the slope. Soils are taxonomically classified into orders, suborders, great groups, subgroups, families, and series in a similar way as are other natural resources are classified. Taxonomical systems are not yet globally unified. This is not the only scientific disadvantage, but it is a practical drawback. We have to consider the fact that soils with their crop and animal productivity are an important basis of the non-predictable daily changing market values of national economic resources.

Soils are as vulnerable as living organisms. As the environmental system changed during the geological evolution of our planet and beginning at the time when living organisms occupied the land, soils developed and changed continually. We can find sometimes the remnants of those past soils. They are called paleosols and we devote more attention to them later on in Sect. 13.3, Granny Soils. These earlier existing soils are sometimes buried under the dust and ash of the past volcano eruption, or they are hidden under thick layers of loess blown by windstorms occurring in the very cool climate of glacials during the last two million years of Pleistocene. Similar to a snowdrift, a loessdrift could form to a depth of several meters, but the time for its development differed greatly from that of snow. Without interruption, loessdrifts were formed during tens of thousands of years. In this way the paleosol, being older and lying under the loess, was preserved. In some instances, paleosols are found hidden below the sediments of rivers.

Abrupt changes of the global environmental system have been caused by the catastrophic impacts of asteroids or a comet colliding with the Earth. Soils were completely destroyed – only small remnants have been found under layers of fossil dust and ash. When such catastrophes happened, soil disappeared from the entire continents. And after such events, because the climate and vegetation changed, the new slowly formed soils differed from the earlier soils. Without the impact of an asteroid, serious soil damage in the last 11,700 years during the Holocene has been caused by human activity. This damaging activity, strengthened during the previous two centuries, has recently led to catastrophic consequences in some regions.

In one of civilization's cradles, Mesopotamia, some dynasties collapsed after introducing a primitive irrigation system that accelerated rather than restricted soil salinization processes typical for the arid region. Crop yields dropped so much that the governing centers were obliged to abandon regions where initially fertile soils gradually became infertile and unproductive. The king and his administration had to initiate a new administrative center in another area of soils with no or little salinity. The new region appeared promising to feed the population.

Great damages are caused by soil erosion by the surface water flow, especially when the physical quality of the soil on the surface is strongly reduced. Water flowing on the slope of bad-quality soil transports soil particles so intensively that the most fertile soil layer, the surface horizon, is smashed. With all of the soil particles being carried away and gullies formed and subsequently washed out, the soil is finally destroyed.

Another example of soil destruction often occurs when the same crop, e.g., wheat or cotton, is grown year after year on the same field. This practice of monoculture, already mentioned above relative to plant nutrient availability and soil biodiversity, also causes deterioration of the strength and stability of the topsoil. Beneficial small lumps of soil known as aggregates disintegrate into separated sand, silt, and clay particles that are detached from the soil surface and carried away by wind forming dust storms. As the wind subsides and particles settle down, their hot dust scorches and destroys crop plants in the wind-eroded field as well as those in nearby fields.

The destruction of agricultural landscapes by severe dust storms during previous centuries is well documented by scientific evidence. The dust storms remembered most frequently are those that destroyed large regions in Ukraine of czarist Russia in the past. Catastrophic droughts accompanying the dust storms were described also in the classical Russian literature of the nineteenth century. Similar dust storms and droughts impoverished tens of thousands of American farmers in the 1930s of the twentieth century. The destructive erosion process, either by wind or by water, or by combination of both may reach so far that all soil is carried away leaving only a completely unfertile sublayer and rough rock.

Soil has a miraculous ability to accept many strange materials and to transform them into an integrated part of its existence. However, if people do not consider the limits of those capabilities and overload a soil with mineral or organic wastes, the beneficial transformation of such wastes into a soil process is no longer a possibility. With such human negligence, those wastes buried in soil eventually become unwanted toxic products that gradually overwhelm the soil into a medium that does not support plant communities.

Modern contemporary society has a new perfect tool for the complete destruction of soils: constructions. We are not speaking about construction of new houses and dwellings for still increasing numbers of population. We are speaking about one- or two-storied shopping centers, warehouses and administration buildings, roads, and airports. They occupy hundreds of thousands of square kilometers where the soil was dug out and replaced by concrete, pavement, and asphalt. With the new surfaces being impermeable and not allowing a drop of rainwater to penetrate into and down through the remaining subsoil, hydrologic cycles are destroyed. Without soil, the natural liquidators of wastes, soil microorganisms, do not grow on these constructed areas and, hence, drastically redistribute the location at which dead organic materials and solid refuse are decomposed. Moreover, the constructed areas become large islands characterized by increased temperature and reduced humidity. Hence, Earth's skin is being attacked daily and continuously shattered.

Up to now, societies have not found measures to reduce this type of soil annihilation. Perhaps the good will of the global community is lacking because of misunderstanding or a lack of reliable information. Commercial developers and their entrepreneurs are still persuading the public only about the short-term convenient and economical aspects of this modern progression of construction. The media repeats, supports, and even glorifies their statements instead of agitating that the *developers* are in reality the *destroyers*. Both terms start with "de" having nothing in common with the French names of aristocrats.

Chapter 4
The Smallest Zoo and Botanical Microgardens

Across any of the diverse landscapes of Earth's continents, we cannot find the extensive assortment of life forms that exists within any one soil. Using the language of science, we say that the greatest *biodiversity* within any region of the Earth can always be found in soils. Scientifically, *edaphon* refers to everything that is alive in any soil. The word is derived from Greek *edaphos* meaning soil and *on* which is analogous to plankton. When we classify edaphon in the simplest way, we list *phytoedaphon* and *zooedaphon*. The prefix *phyto* was derived from the Greek *phuto* meaning plant or something that has grown. Subsequently, the word appeared as *phyto* when transcribed into scientific Latin. The Greek *zoon* is a living creature or an animal.

The organisms comprising two components of edaphon transform all organic compounds in soil. Their quantity and species composition are responsible for the intensity of all transformation processes including the weathering of rocks. Their "home" is one or more soil pores, i.e., the space between soil particles not occupied by organic matter. Soil pores are either completely or only partially filled by water. If the pores are full of water, the soil is waterlogged, and without any air in its pores, the soil suffers a lack of oxygen. Processes proceeding without oxygen are called anaerobic. The literary translation is "without air"; the Greek *aer* is air, or the content of lower atmosphere that we breathe; roughly the same is *aeros*. The Greek prefix *an* means no or absence of. Aerobic processes occur when pores partly filled by water also contain air with enough oxygen for aerobic conditions.

Because of their microscopic and even much smaller sizes, the great majority of edaphon is not visible to the naked eye. Without ever seeing such infinitesimal organisms or witnessing the minuscule domains in which they induce aerobic and anaerobic processes, the lay public frequently underestimates or is completely unaware of the irreplaceable role of edaphon. Without edaphon, soils cannot develop or exist.

© Springer Science+Business Media Dordrecht 2015
M. Kutílek, D.R. Nielsen, *Soil: The Skin of the Planet Earth*,
DOI 10.1007/978-94-017-9789-4_4

4.1   Viruses

The smallest organisms living in soils are viruses. Being strictly parasites, they live and reproduce mainly in bacteria and generally in phyto- and zooedaphon. The most important kinds of viruses in soils are those living in bacteria cells. They are called bacteriophages or simply, phages. Their name carries the Greek *phagein* meaning to devour or to eat. These phages are able to eliminate some populations of bacteria because they "eat" the bacterial genome and liquidate the hereditary information required for bacteria to reproduce.

The negative impact of phages attacking nitrogen-fixing bacteria living symbiotically with roots of legume crops is well known by farmers. They know that alfalfas growing in their fields in the absence of symbiotic cooperation with bacteria do not absorb enough nitrogen to produce economic yields.

4.2   Bacteria

The next smallest soil microorganisms are bacteria – single-cell creatures having diameters or lengths between 0.2 and 2 μm. They belong to prokaryotes, cells without a nucleus, with a name derived from the Greek words *pro* meaning before and *karyon* meaning kernel or nut. Without a nucleus, bacteria lack mitochondria to serve as an internal cellular power or energy source. Mitochondria could be compared to the power plant of the cell, but they have other more important functions for the life of the cell. For millions of years prokaryotes were the only form of life on Earth until eukaryotes evolved. A eukaryote is an organism containing mitochondria and other complex structures. The Greek prefix *eu* means good. Eukaryotes include animals, plants, and fungi.

Bacteria are present in soils mainly in the form of a thin biological film covering the solid particles. Without any chlorophyll within their bodies, they cannot use solar radiation as a source of energy needed during their lives. For an alternative energy source, they take over the energy forming the bonds between atoms in the organic molecules within dead organisms and excrements. When these bonds are broken, energy is released and further used for their life processes. This decomposition of the organic molecules sometimes produces nitrogen in the chemical forms of nitrite and nitrate. The process is called *nitrification*. Completing the two-step process to nitrate is important to avoid the toxic influence of nitrites upon plants. Another decomposition product of organic molecules is ammonia and the process is denoted as *ammonification*. It typically occurs with inadequate levels of oxygen, i.e., under anaerobic conditions. However, ammonification also happens during initial stages of decomposition and even simultaneously with nitrification. Whenever the two processes occur together, nitrification usually dominates provided there are sufficient amounts of oxygen in the soil pores (Fig. 4.1).

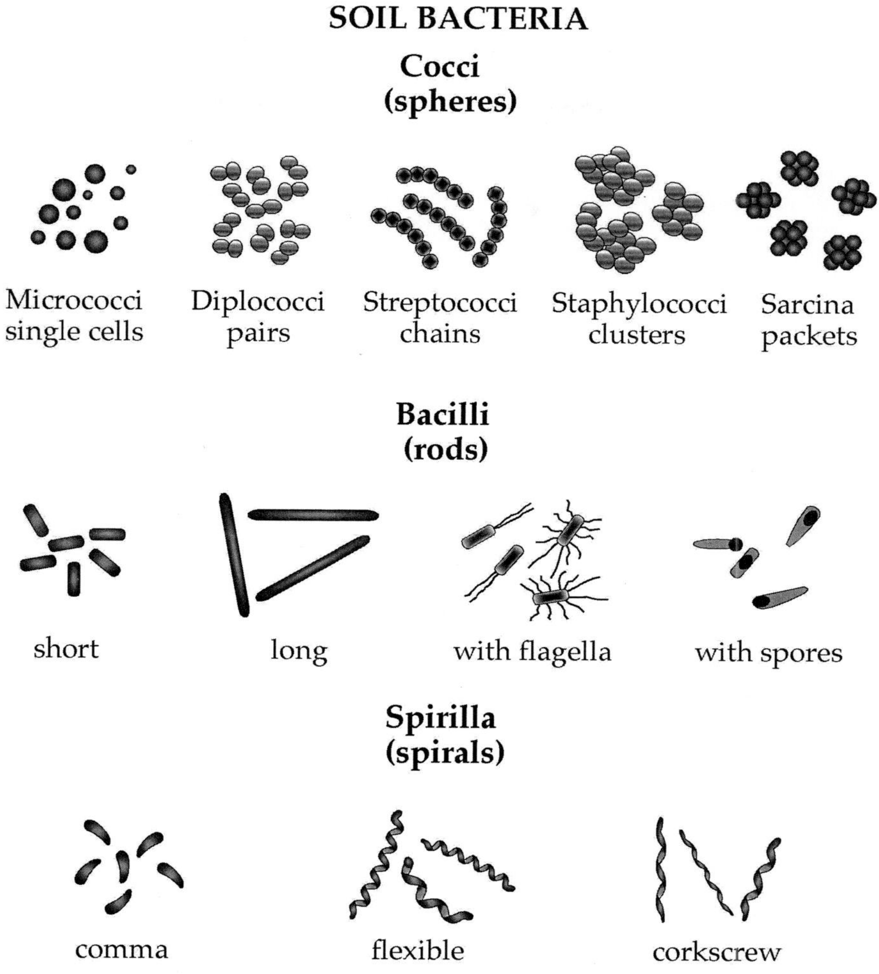

SOIL BACTERIA
Cocci
(spheres)

| Micrococci single cells | Diplococci pairs | Streptococci chains | Staphylococci clusters | Sarcina packets |

Bacilli
(rods)

short   long   with flagella   with spores

Spirilla
(spirals)

comma   flexible   corkscrew

Fig. 4.1 Soil bacteria

Each of the many steps of decomposition of organic molecules is produced by different specific families of bacteria, all of which are known as chemotrophs that derive their energy from chemical reactions. Chemical reactions (known as an oxidation process) under aerobic conditions are often 10 times faster than those (known as a reduction process) under anaerobic conditions. Bacteria are active in many oxidation and reduction processes in soils. Speaking in a very simplified chemical language, oxidation occurs when the outside shell of an atom loses an electron and reduction occurs when the outside shell of an atom gains an electron. Bivalent iron compounds are changed to compounds containing trivalent iron during oxidation. In other words, the number of electrons missing in each atom of iron increased from two to three. The opposite process is reduction when the trivalent iron oxides are changed to bivalent iron oxides. Analogously, compounds containing tetravalent manganese are transformed by reduction to those containing trivalent manganese in anaerobic conditions that exist in waterlogged soils.

The type of transformation of soil organic compounds depends upon the activity of families of bacteria related either to electron donation (oxidation) or consumption (reduction). Aerobic bacteria assimilate up to 10 % of the carbon, while anaerobic bacteria assimilate only about 0.5 % or less, leaving behind many waste carbon compounds. Such assimilation has many practical consequences, e.g., up to one-third of the fertilizer applied to a farmer's field to provide nitrate for his crop can be removed and lost from the soil by microbial respiration. Or, when aromatic compounds are applied to soils to wipe out insects, they also serve as an instantaneous source of organic carbon that stimulates the growth of bacterial families. As a result, the insecticide is quickly demolished instead of the insects. This unwanted oxidation occurs when phenol and benzoic acid are used as insecticides.

Another example is the existence of sulfate-reducing bacteria living in the anaerobic soil of a rice field soil on the roots of the farmer's crop. The action mentioned here of another variety of soil bacteria capable of reducing the structural trivalent iron locked inside of clay minerals is discussed in more detail in Sect. 5.2.2. There are many, many more examples of bacteria transforming never-ending numbers of mineral and organic components of soils.

In order to make at least one of our examples more transparent, we compare the trivalent iron in crystal lattice of clay minerals to a prisoner with life sentence. Dressed in a red coat, he has been confined to an isolated island in a prison surrounded by concrete walls and a fence of electrically charged wires. All guards believe that his escape is impossible. In addition to prisoners wearing red coats who have many restrictions, there are prisoners wearing blue coats living with no restrictions. They are bivalent iron ions. Once, a couple of boats paddle noiselessly to the prison by a bacterial crew trained how to smuggle a blue coat into the prison. They succeed and the life-sentenced trivalent iron receives and immediately puts on the blue coat. He is thus changed to a prisoner without restrictions. The mode of smuggling is the secret of reduction bacteria; they inherited it after millions of years of attempts.

The number of bacteria living in the soil is far beyond our imagination. About one hundred trillion bacteria live in a column of soil having a surface area of 1 m^2. That number of bacteria could remarkably increase by a hundred times. Since they are very small, it is instructive to say that if we were able to sieve all bacteria from 1 g of soil, we would count up to one hundred million bacteria. In other words, in a handful of soil there exists the same number of bacteria as the number of people inhabiting our planet Earth.

A very important group of soil bacteria are *actinomycetes*. In many species when observed under a microscope their shape resembles that of fungi – their elongated cell is extended by very fine filaments. Although the majority of actinomycetes are aerobic, a few are anaerobic. They get their energy from the decomposition of organic materials. The process leads to the release of plant nutrients from the decaying remains of plants and to the formation of humus (see Sect. 5.3 later on). Actinomycetes are therefore an important crucial link in the carbon cycle. Increasing temperature in soils intensifies the rate of decomposition. Thus, climatic warming has a positive influence upon their processes and has a direct bearing on the release of higher plant nutrients and the increase of soil productivity, provided organic materials are adequate to supply all of the plant nutrient requirements. The warming results in increased organic matter decomposition, faster release of plant nutrients, and greater amounts of vegetative life, all of which accelerate releases of CO_2 into the atmosphere where it is quickly utilized by photosynthesis to increase the generation of plant bodies which, in turn, provide increasing amounts of plant remains to eventually decay. Climate warming therefore accelerates the carbon cycle.

And vice versa, climate cooling leads to the decrease of actinomycetes' activity and to slower decomposition rates of plant remains. This is why low levels of agronomic productivity characterize tundra soils of cold regions of Earth, even though they are relatively rich in organic materials.

Many of the aerobic actinomycetes produce a specific odor of moistened soils, e.g., after a rain. Experienced farmers recognize different levels of soil fertility according to this odor when they wet a handful of soil. After stirring the soil in their palm or between their fingers, the occurrence of strong pleasant odor means that the potential fertility of the soil is high. The soil scientist's explanation is simple. The procedure and knowledge gained by the farmer is due to the strong immediate response of an abundance of actinomycetes decomposing decayed organic matter (most likely plant residues) and releasing plant nutrients with the final consequence of the production of a pleasant odor. Everything looks simple if we and the farmer have a lot of experience.

4.3 Soil Fungi

Soil fungi are members of a large group of eukaryotic organisms and are the most numerous constituent in the phytoedaphon. They are scientifically classified as a kingdom among kingdoms of animals, plants, etc., in the domain of Eukaryota. Their size is about ten times greater than the size of soil bacteria. They contain more than one cell and individual fungi are usually linked into a long chain of micromolds called mycelium consisting of a mass of branching, thread-like hyphae of few micrometers thickness. Their numbers in topsoil are very high, in 1 g of soil there is as many as one million fungi. On 1 m² of topsoil we find about one million fungi. They are most widely present in the acidic conditions of peat and forest soils where they may be even more numerous than bacteria.

Because fungi contain no chlorophyll, they must gain energy for their life processes from the chemical breaking of bonds in organic compounds. They use therefore chemical sources of energy rather than that of direct sunlight. This is why they have to participate in the decomposition of fresh organic matter like lignin and cellulose. Hence, decomposing straw very efficiently, some species of fungi produce special hormones important for the growth of roots. Other species cooperate to produce antibiotics important in the protection of plant roots against diseases. On the other hand, some cause diseases and therefore belong into various groups of plant pathogens. Still other species live symbiotically with plants roots – a relationship known as mycorrhiza. The name is derived from two Greek words mykos meaning fungus and riza meaning roots.

Mycorrhiza enables fungi the access to simple organic compounds like glucose and sugar from the roots of plants. For fungi these gained compounds are simply decomposed and the energy important for fungi is gained very quickly and effectively. In return, the long and dense net of ultrafine hairlike mycorrhizal mycelia of fungi enables the plants to gain nutrients and water from a substantially greater volume of soil than that reachable by plant roots. It is not merely a simple physical difference of soil volume. The more important aspect is that the permeability of fungal cell membranes is much greater than the permeability of plant root membranes. Hence, nutrients enter the plants simultaneously along two paths – one directly from soil to the root and the other indirectly into the root after first being efficiently absorbed by the dense net of ultrafine mycorrhizal mycelia surrounding and in the immediate vicinity of the plant roots. Moreover, unique chemical reactions occur within mycorrhizal domains that lead to more efficient release of plant nutrients from the original dead organic materials. Mycorrhizae are especially beneficial for the plant partner in nutrient-poor soils and when the soils are revitalized after damaging technology, as, e.g., the mining of brown coal in the immediate vicinity of a soil surface.

4.4   Soil Enzymes

Soil enzymes are powerful products of soil bacteria and fungi activities during which bacteria usually play a leading role. Enzymes are proteins produced inside the cell and exported out of the cell into the soil solution where they regulate the breaking down of the complicated structure of the dead remnants of plants and animals that is commonly called a substrate. The processes lead either to the breakdown of organic matter only or to the release of plant-accessible forms of nutrients. The latter process is commonly known as nutrient mineralization.

Because enzymes are energetically costly for the cells to make, they are tightly regulated and made only when absolutely necessary. Regardless of substrate availability, the cell initially begins the production and expulsion of tiny amounts of enzyme as a mechanism to detect the existence and nature of its substrate. If a substrate is present, these constitutive enzymes generate signals that induce additional enzyme synthesis. Although the synthesis of enzymes is sensitive to the presence or absence of many other factors, the most important is the presence of sufficient amounts of organic substrate. Hence, we are obliged to keep the organic content of substrate at or above threshold levels by green farming, introduction of cover crops, and other steps leading to conservation of organic materials in soils. The quality and quantity of enzymes respond to soil management changes long before other soil quality indicator changes are detectable. The detection of various types of enzymes and their relative amounts is a practical and objective tool for classifying soil quality.

4.5 Algae

Algae differ from the micro-edaphon mainly by their existence either as single-cell or multicell organisms and by their ability to convert the radiation of the sun into the chemical energy utilized in all their life processes. They exist at soil depths repeatedly penetrated by rays of sunlight with their maximum numbers occurring at or very close to the soil surface. Their numbers vary between 100 and 3 million per 1 g of soil. In other words, on a land area of 1 ha, the soil is endowed with 0.2–1.5 tons of algae. Sometimes they are distinctly visible as a greenish or bluish film on top of the soil surface. With water being essential for their existence, families of algae are categorized on the basis of optimal or specific ranges of soil water content. Algae are known as the pioneers who start the weathering of rock surfaces, often in a symbiotic relationship with fungi and lichens. They produce various slimes by which individual soil particles are glued together into bigger aggregates which resist being destroyed by water, raindrops, or water flowing on the soil surface after the melting of snow, a heavy rain, or an intentional irrigation.

Through photosynthesis, soil algae facilitate the aeration and oxygenation of both submerged and un-submerged soil environments by liberating large quantities of oxygen directly into the soil profile. In un-cropped soils, they are especially helpful checking the loss of nitrates through leaching and drainage.

4.6 Zooedaphon

Microfauna are the most abundant portion of the soil animal kingdom. They are dominantly protozoa, and in addition to them there are also rotifers and in lesser extent tiny nematodes. Their sizes range roughly between 2 μm and 0.2 mm, but they can grow as large as 1 mm. Within a column of soil having a surface area of 1 m^2, we find from 10 million up to 10 billion protozoa, and under favorable conditions their mass may reach up to 100 g/m^2. They gain their life energy not just from the decomposition of plant residues and dissolved organic matter in soil water. They absorb food through their cell membranes, and some as amoebae and flagellates also feed on bacteria. They surround food and engulf it. Others have openings similar to mouth pores into which they sweep food. All protozoa digest their food in stomach-like compartments called vacuoles. In this way, they select and rejuvenate the population of soil bacteria. Many parasites that affect human health or the economy of agricultural production of food are flagellates. The mass of nematodes

SOIL FAUNA

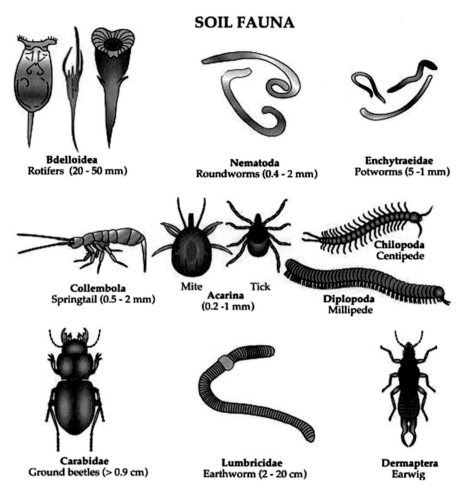

Bdelloidea
Rotifers (20 - 50 mm)

Nematoda
Roundworms (0.4 - 2 mm)

Enchytraeidae
Potworms (5 -1 mm)

Collembola
Springtail (0.5 - 2 mm)

Mite   Tick
Acarina
(0.2 -1 mm)

Chilopoda
Centipede

Diplopoda
Millipede

Carabidae
Ground beetles (> 0.9 cm)

Lumbricidae
Earthworm (2 - 20 cm)

Dermaptera
Earwig

Fig. 4.2 Some of the soil macrofauna

is lower and it happens that it is even by orders of magnitude less than that of protozoa (Fig. 4.2).

Macrofauna are known for their formation and use of new pores in soils in addition to utilizing existing soil pores. These pores enhance two phenomena – soil drainage and the penetration of atmospheric air into soils – that improve all soil oxidation processes. While forming such pores, macrofauna play an important role in remixing mineral soil particles with organic compounds that originated partly as the result of the decomposition of dead bodies and partly as the product of vital activity of macrofauna organisms.

Among this group of macrofauna, worms and especially *earthworms* play a dominant role. Today, it is common knowledge that Charles Darwin was the originator of the biological theory of evolution presented in his book, *On the Origin of Species (1859)*, and countless persons also credit him for his description of earthworm activity presented in another book, *The Formation of Vegetable Mould Through the Action of Worms, with Observations on Their Habits (1881)*, based on his experiments in the garden at Down House when he was a young scientist. During those times, the majority of humans thought earthworms were merely pests. His experiments showed that earthworms were beneficial for turning over the soil, tunneling through it while chewing it up and excreting it out, thereby making it more fertile. It is remarkable that during his life, copies of his earthworm book were sold more frequently than his book on evolution. But still more remarkable is the fact that Darwin was not the first to reveal the beneficial activities of earthworms. A century before Darwin's book and even before Darwin's birth, Gilbert White wrote not only statements but a book near the end of the eighteenth century carefully explaining the immensely beneficial functions of earthworms in agriculture. Today, we know that there are about 10–100 thousand pieces of earthworms under a soil surface area of 1 m^2 and that their mass usually ranges between 1 and 50 g/m^2.

Ants also play a special function in soils. They loosen the soil by forming fine corridors and nests, they mix the soil, and they contribute to a regular distribution of dead and decaying parts of edaphon in the mineral soil material. All their activities improve the thermal isolation both inside the anthill and in its large surroundings. They try to maintain an optimal soil water content for the biological processes in the anthill that, in fact, acts positively for the soil-plant relationships in the great majority of cultivated crops. Since ants modify the biochemical processes to the conditions required by the life of the entire ant community, they act in a similar positive way upon soil processes within the anthill.

Recognized as groups of "social" insects, termites are relatively abundant in the tropics and subtropics and are classified into seven families containing about 2,800 species. Recalling social communities of ants with a queen and king, nymphs, workers, and soldiers, termites in spite of their organized behavior are not genetically related to ants. Even so, they are commonly known among Australians as "white ants." Overall, their activities have positive effects – feeding on dead plant material, wood, and soil, they mix the transformed matter together that contributes favorably to the soil environment. But some of them can cause damage to buildings, house equipment, or plantation forests. Termite workers undertake foraging, food storage, and building the nest. They have special gut flora and enzyme productions that enable them to digest cellulose. And here we find why so many Europeans and Americans curse termites after a stay in the tropics. When I (Mirek) returned to my temporary home in Khartoum after enjoying a lovely vacation in Europe, I happily entered my bungalow. Before leaving for Europe, I locked my clean shirts, etc., in a wooden wardrobe. After arriving home from the Khartoum airport, I decided to change into a fresh, clean shirt. To my surprise, when I attempted to merely unlock the wardrobe, I pulled out its metallic lock. While staring and holding the metal in my palm, the rest of the wooden door lost its shape and suddenly fell down as dust. Immediately after that, the entire wooden cabinet disintegrated into a heap of dust intermixed with my perfectly washed and ironed shirts, slacks, and linen.

In Sahel, the dry climatic strip lying between the Sahara desert and the Sudanian savanna, and in the savanna itself, termites construct quasi-cylindrical mounds several meters high with unique thermoregulation that prevents the complex structures inside the mounds from overheating. With this type of air conditioning, termite workers are able to cultivate fungal gardens and harvest sufficient fungi to feed each and every termite in the entire colony.

4.7   Soil as a Microgarden?

In spite of the fact that soil is a biological system with a high microbial diversity, the space occupied by microorganisms is very small, since there are only few sites appropriate for microbial life. When biological molecules such as DNA and enzymes are adsorbed on the surface of particles due to their negative charge, they are resistant to further microbial degradation. As a result, with those parts of the pore space not being suitable for microbial life, microbes are not equally distributed within all soil pores. In those portions of pore space providing free access to food for microbes, we find many thousand times more microbes than the average number of microbes in one cubic centimeter of soil. A comparison of the distribution of micro-life in soil pores to that of cultivated plants in a regular garden is not appropriate because we expect that plants grow across each and every square foot of the garden, while we detect absolutely no microbes occupying micro- and nano-spaces within soil pores. It is more appropriate to restrict the comparison to a big botanical garden and a zoo where on an area of a couple hectares we encounter plants of several climatic zones and animals from polar bears to leopards. This comparison is more realistic because we are familiar with the different kinds of species potentially living together in a garden and a zoo. Indeed, number of species in only 1 cm³ anywhere within the garden is orders of magnitude higher than the number of animal species in the entire zoo. Botanical gardens and zoos also have spaces without any obvious plants and animals. They are paths and roads reserved for visitors and service facilities for maintaining the garden and zoo. We emphasize here a certain similarity to the irregular distribution of microorganisms in soil pores, while a great biodiversity also exists in the soil as well as in the botanical garden and the zoo.

The soil pore space also has other specific features. It encloses fine roots and provides the structural porosity for the rhizosphere. It is a nutrient-rich environment that tends to partially curtail the impact of the abovementioned spatially variable distribution of microbes. Root exudates offer food to all microbes within the entire pore space and especially to those on the surface of fine roots. The numerous interactions between a plant and microorganisms in its root zone are the reliance of the microorganisms to utilize root exudates and to interact or compete with other types of microorganisms. The interaction between plant and soil microorganisms may range from mutual profit up to parasitism.

Chapter 5
The Birth of Soils

Readers, at this moment, we beg each of you not to get discouraged as you wade through this chapter, the longest one in our book, trying to grasp all of the essential facets of the birth of any soil. To explain and interconnect each of them required elucidation of a network of physical, chemical, and biological processes occurring together within specific domains below and above the Earth's surface during various time intervals ranging between microseconds and thousands of years. Sit tight while you read and perhaps reread parts of it before we proceed together into the next chapter.

Soils are never born all of a sudden like a woman delivers a baby or cow delivers a calf, nor is the origin of a soil a relatively rapid process like the germination of a plant from a moist seed. At this point, our inclination to compare soils with living organisms is limping along in an unconvincing manner. Actually, the birth of a soil begins with gradual transformations of its originating materials, usually compact rocks or their weathering products that vary in size from big stones to dust particles. During the time of floods, the smaller particles are transported by water mainly to a neighborhood of rivers or by rivers into an ocean where they settle down to initially form a muddy riverbed or seabed. After the flood, the muddy material gradually becomes dried out, compacted, and chemically fixed to eventually form a solid, sedimentary rock – after which, the process of weathering starts all over again. Sometimes the particles are transported by wind during dust storms. The transported material could be enriched by decayed plant residuals that glue the small mineral dust particles together. Much less frequently, whenever excessive amounts of plant residues prevail in stagnant water enriched with small mineral particles, another member of the family of soils, peat, makes its appearance.

Combinations of many individual reactions running at the same time and influencing each other create the initial transformation of the dense mineral matter. Moreover, the rates of reactions increase or decrease with time. Since they are triggered by outside factors and since those factors are frequently not constant in time, the weathering rates of rocks during any given time period may be consistently high or low as well as remarkably stable or variable during other periods. The same is

© Springer Science+Business Media Dordrecht 2015
M. Kutílek, D.R. Nielsen, *Soil: The Skin of the Planet Earth*,
DOI 10.1007/978-94-017-9789-4_5

valid for all other processes leading to the creation of soils. Inasmuch as transformations of various components of organic matter coincide with most of these mentioned processes, intermediate and subsequent organic products and by-products
depend upon a complex series of reaction rates. When we compare the complicated
process of creating a soil that takes decades to thousands of years with human or
animal life, we recognize that a soil's existence passes through childhood, adolescence, and adulthood. However, it is more complicated. When external climatic
conditions change, soils may return to their adolescence or even to their childhood
in order to find an alternative path of evolution and aging. And in some instances,
they may even die – akin to the final event of human or animal life. Soil formation
is a very complicated process lasting for a long time and keeping its own dynamics.
The complex system allows substances to enter and leave, while concomitantly
encouraging the acceptance and release of specific energies required for critical
soil-forming processes.

Genuine soils first originated about 400–350 million years ago on sediments
along the mouths of rivers flowing into big lakes and into oceans. These wet porous
media enabled the entrance of primitive plants from the shallow sea onto land. Even
though the sediments were not fixed owing to the absence of plants and easily
washed and blown away into rivers and oceans, voluminous amounts of weathered
rock materials were available for the birth of soils.

A new situation began when the first pioneering plants found their feet in soils.
With their roots actively aggregating the originally loose sand, silt, and clay materials, plants protected the sediments and developing aggregates from the continuous,
relentless splash of water that previously washed them down into rivers, lakes, and
oceans. The plants also reduced the blowing force of strong winds and tornadoes.
When the plants died, their dead remains provided perfect nourishment for the
microorganisms already present that were busy taking the energy released as they
disturbed the chemical bonds in rocks and minerals. Other troops of microorganisms were moving from waters of oceans and rivers ashore into the pores of sediments, driven by the chance of finding residues of plants and nutritional ingredients
in excess. However, because not all pores were filled completely with water, they
had to accustom themselves to their new environment. As a result, new families of
microorganisms evolved as they adapted to the flip-flop tactics of water and air –
that is, the solid surfaces of pores on which they resided were sometimes completely
surrounded with free liquid water, and at other times, the microorganisms were
touched by varying amounts of air that partially filled the pores. This variation of
both air and water was the situation when nearly sterile mineral sediments started to
attract colonies of microorganisms and subsequently attracted emerging plants.
Such sediments serving as a home for microorganisms that modified local physical
and chemical properties also became a home for plants. The higher the density of
microorganisms, the more attractive was the newly formed home for plants. This
home was the birth of new global environment we now call soil. Soils first originated near the water of rivers and seas. Then in the run of million years, the soil-
forming processes were extending together with pioneer plants into regions more
distant from rivers and oceans. Ultimately, soil occupied all terrestrial regions of

continents and islands. And, as we show two pages later on, the configuration of continents was never constant in the past. Supercontinents existing in early stages of the Earth's history broke up into continents and islands which later rejoined at least partly into other continents only to break up and partially join together again. The wandering of their ever-changing positions, shapes, and sizes was so fascinating that the South African geologist Alexander du Toit wrote in 1937 the book *Our Wandering Continents*. This geological unrest lasting for hundreds of millions of years influenced soil genesis a countless number of times.

The pioneer plants mentioned above acquired a new habit of tremendous global importance – they gained their energy directly from solar radiation, extending what eukaryotic microorganisms and marine algae were doing earlier, but with smaller impact upon the Earth system when we compare it to the role of terrestrial plants. With the solar energy the plants transformed water and atmospheric CO_2 into organic compounds. They also required nutrients such as phosphorus, potassium, and others that became available during the occurrence of chemical weathering of minerals composing the original rocks. Upon being absorbed by plant roots, those nutrients were distributed into all organs of the plant's body. The carbon cycle originated from the same sequence. Carbon as a simple gas CO_2 is accepted from the atmosphere and transformed into more complicated organic compounds that provide nourishment for terrestrial animals and alimentation for microorganisms. In both cases, this part of the C cycle is closed inasmuch as they are again decomposed to CO_2. If we intend to balance the whole cycle, we must include within the cycle the carbon kept in bodies of animals and in their secretions. Many of the compounds formed in both processes of composition and decomposition have decisive influence upon soil physical and chemical properties. The most remarkable characteristic of all initial processes of soil formation was the presence of living organisms, since without them soils could not originate and nowadays they cannot exist. Just super-briefly: no terrestrial life – no soil and vice versa: no soil – no life on continents and islands.

The mineral material from which soil originates is called parent rock. It could be a solid rock, or a silty sediment called loess blown by winds tens of thousands of years ago and deposited, or sediments transported and deposited by rivers, or large areas of sea bottom originally muddy-like and dried out when the seawater level sank by tens of meters. Rocks have their own history of origin. Since it is not our intention to offer a complete family tree starting at the origin of our planet Earth and since this detailed knowledge is not required in soil science, we start with the solidification of the melted magma on or close to the Earth's surface. When the hot lava is cooling and solidifies, a strange transformation occurs. The hot liquid or plastic material appearing rather like a homogeneous corn mush solidifies into a mixture of small or bigger minerals fixed firmly together. There are two types of rocks differing one from the other because of the size of their minerals. The first example is granite with whitish, gray, and black minerals that can be recognized and seen by the naked eye. The second example is basalt that looks like a dark gray to black monolithic stone and only a more detailed study optimally made with the aid of a magnifying glass will reveal individual minerals having different colors. Compared with granite, white dots representing quartz are missing in basalt. Basalt prevails in the Earth's crust. Having their origin deep within the Earth, igneous rocks are classified and grouped together. If they are subsequently subjected to increased pressure and heat, they lose their original form and appear within the group called metamorphic rocks.

Rocks are usually stable unless their external pressure and temperature conditions change as a result of being shifted vertically. These changes have the consequence of decreasing mechanical and chemical stability if the rock is lifted up. On the other hand, the submersion of the rocks into great depths is accompanied by heating and by an increase of pressure. The rock starts to be melted and gains the appearance of a very dense pulp or paste. Extremely great changes of pressure and temperature occur when lithospheric plates hit one another or when such a plate is broken. A lithospheric plate is the scientific designation for a piece of the Earth's fractured crust. Lithospheric comes from the *Greek lithos* meaning stone or rock and *sphaira* meaning sphere. These plates are also known as tectonic plates. Tectonic is derived from the Greek *tekton* which means builder. Today, we recognize that continents are huge lithospheric plates slowly and continually floating on liquid magma kilometers below them. Several times in the past, these plates were connected forming a single supercontinent that later broke apart. Subsequently, some of the remaining plates collided to form and build waves with their tops becoming mountains. The hypothesis about this continental drift appeared at the end of the sixteenth century and was theoretically explained in the twentieth century by Alfred Wegener. A meteorologist by university education, he actively participated and conducted meteorological expeditions in high levels of the atmosphere using weather balloons and in the extreme polar climate. Incidentally, in 1906, he together his brother on the occasion of taking meteorological data set world records in the height and in the continuous length of a balloon flight. He was also the first scientist made famous by overwintering on the inland Greenland ice sheet where he organized the first boring of ice cores in addition to taking meteorological data. He was exceptional in combining adventurous expeditions with utilization of his own measured data in theoretic developments in many domains of geophysics – among them is the formulation of the first continental drift theory that was not fully accepted until after 40 years when it was experimentally confirmed and broadly accepted in the 1950s. Many new measuring procedures provided support for continental drift theory and for the model of plate tectonics. The model contributes to the explanation of many earlier not well-understood discoveries, e.g., why recent deserts like the Sahara did not exist several million years ago.

Locally some parts of the Earth's broken crust were pushed downward and heated. This was the situation when the initially igneous rocks gained their metamorphic features. Gneiss, a typical example of these metamorphized rocks, is found in abundant quantities as a major part of the Earth's lower crust.

When the rocks are formed from external materials like sands or seawater precipitates, we denote them as sedimentary rocks. An example of the first group is sandstone. The second group is represented by limestone that may contain plant residuals and shells of small mollusks buried millions of years ago. Again, if increased pressure and heat act upon them, the rocks are transformed into metamorphic rocks analogously to igneous rock transformation. Slates and marble are examples. The great majority of the top upper part of the Earth's crust is formed by sedimentary rocks that originated from transported products of weathering. On the other hand, a minority of the weathered material originated in place without transport. Owing to a combination of physical, chemical, and biological processes, sedimentary rocks were consolidated and bound together by various binding agents such as compounds of ferric iron or lime (calcium oxide) manifesting, respectively, colors of rust or light beige.

5.1   Rocks' Weathering

The word weathering indicates that the parent materials, i.e., rocks, are being transformed under the influence of weather. Initially, after the entire extent of each solid rock is cracked or partially crushed, water enters the cracks, and after its long-lasting direct contact with the rock's minerals, their chemical composition is changed. The dissolved simple compounds associated with the macerating rocks are transported away by water leaving a residual of pulverized rocks and transformed minerals.

Weathering processes are most intensive at the surface of the Earth where it is in direct contact with the atmosphere and hydrosphere. The intensity decreases with depth below the surface and depends mainly upon the strength and time span of numerous factors. The depth of weathering ranges from several millimeters to tens of meters. Some rocks weather easily, while others are very resistant. For example, sedimentary rocks weather rapidly to unconsolidated smaller particles of sand and silt, while volcanic rocks are much more resistant to weathering. One should not make the mistake of thinking that individual particles of sand and silt are at their final stage of weathering. Although the weathering of solid rock to small individual particles takes many, many years, further sequences of weathering continue to proceed at various rates without ever reaching a true finality characterized by well-defined end products. Intensities of weathering are influenced by the mineralogy of individual particles and their layering. It is common to believe that rates of weathering typically decrease with time – *high* at the start and *later* diminish to a legendary rate of zero. Such a belief is a myth because words like *start* and *later* have imprecise meanings when the time scale of geologic processes is not clearly defined or understood. At any time, weathering intensity depends upon the immediate condition and extent of disruption of the rock and upon the characteristics of the climate. In the warm, humid climate of the tropics, the intensity rate is high compared with that in temperate zones.

5.1.1   Physical Weathering

We are going to start the description of weathering with *physical weathering* that looks like the simplest among all soil-forming processes. When a mineralogical heterogeneous rock surface is radiated by the sun, the dark minerals absorb more radiation than the light-colored minerals that strongly reflect the incoming radiation. We know it from our own experience since we dress in whitish clothing during summer time in order to avoid our own overheating. The minerals composing the rock increase their volume when their temperature rises. As the dark minerals are warmer than the light minerals due to the absorption of sun radiation, they have the tendency to increase their volume more than whitish minerals. We have to consider also the fact that each mineral has its own value of volume increase with a specific rise of temperature. Accordingly, we know that each mineral has its own coefficient of volume extension. However, the compactness of the rock prevents the irregularities in volume extension to be realized and causes an internal stress. During the relatively cold night, the volume also changes, but in the opposite direction. With the minerals having the tendency to shrink, an internal stress is again produced. When the tendencies are many times repeated, the individual minerals start to be separated and cause cracks in the surface of the initially compact rock. The cracks are microscopic, not visible by the naked eye. Sometimes the process continues to greater and greater depth, while the individual minerals keep their original position. We have observed this cracking in the dry climate of deserts and semideserts. We described the occasion of discovering such a product of physical weathering in our book, *Facts About Global Warming: Rational or Emotional Issue? –* Catena, 2010:

> I (the first author, MK) took my students of the Khartoum University to an excursion to the north to visit the landscape of the sixth Nile cataract. The students sat on the bed of the truck while I sat next to the local driver with one of the students as an interpreter. During the first half-hour we drove in the direction of rutty wheel tracks. When there were no more tracks or any obvious trail to follow, we continued to take an unmarked route roughly to the north, which only the driver knew that eventually led to the Nile cataract. The driver's name, translated from Arabic, was Crocodile. Since I knew that animals have a better instinct for direction, I trusted the driver mainly because of his name from the animal kingdom. The students and I had hiking boots while the driver Crocodile was wearing homemade sandals with soles cut from an old tire. I decided to stop close to an apparently unweathered coarse granite rock. Knowing that visual teaching is the most effective, I walked to the rock where I swung my right foot to the rear and then I kicked with all my force as if I were kicking a penalty during a football game. The students were silent for a second and then they exploded in a cheerful ovation when they saw that my entire boot penetrated into the "rock" like in a heap of sand. Everybody tried the kick, even the driver Crocodile. He did it with less force – his foot was in a sandal. My explanation on physical weathering of coarse-grained rocks followed in English. I also did not forget to mention that fine-grained rocks do not all weather in the same way. In the dry hot climate, they are sometimes covered by a compact glassy coating as we had the chance to see later on at the Nile cataract. All of my "lecture" was in English. With the driver knowing only a couple of English words, he definitely never heard nor even understood terms like weathering and glassy compact coating.
>
> We continued on our drive to the cataract where the Nile valley narrowed and the stream was very swift and filled with rapids as its water flowed with difficulty around the obstacle of large, fine-grained igneous rocks of Sabaloka mountains. Here we stopped again, and

even before I could say a single word through the interpreter, the driver Crocodile, smiling with pride, quickly ran to the rock to be the first to kick it. He kicked with all his might, but the solid glassy rock did not budge and left his foot bleeding and swelling with excruciating pain. I had to drive on our return to the university, with the driver Crocodile only being able to navigate me with the help of my student interpreter. I am absolutely confident that those students kept in mind for the rest of their entire lives that vivid lesson on physical weathering, and on the role of climate, which is modified by the specific nature of weathering rock. And the driver Crocodile is still probably thinking that the white "khavadia" is not worth trusting.

In regions outside of deserts, there are zones with rainy seasons where water is filling fine cracks freshly appearing on the surface of otherwise compact rocks. If it happens in zones of a mild climate, frosty days and weeks having temperatures below freezing are prevalent during the winter. Water in the cracks freezes, and since ice has a bigger volume, the freezing of water acts like a chisel expanding the original thin cracks. As the freezing, or combination of freezing-melting-freezing repeats for many years, the individual minerals are finally released. A similar but less intensive effect transpires when clay and soluble products of chemical weathering fill the cracks. Water running across the mineral surface after a thaw or rain loosens and transports unbound minerals away. As the suspended liberated particles hit each other as well as other obstacles, they are scraped and abraded. When they finally subside, they are rounded and free of their original shape.

The fine roots of plants act in a similar way when they penetrate into tiny cracks where they are searching for water and nutrients released from the rock by chemical weathering. The fine hairlike roots are the extension of bigger roots, and both are growing to produce pressures that split microscopic cracks and pores. The dead roots remain in cracks, and as they are decomposed and chemically transformed, additional stresses begin.

A special type of physical weathering is the action of glaciers. Glacial ice flows at a very slow pace. The force driving the ice is the resultant of the continual force of gravity being linked to the variable action of other forces. Rather than describing the physics of ice flow, we accept the fact that the driving force exists and glaciers move slowly and continuously. The slowly moving ice causes abrasion of the rock at the front, bottom, and sides of the glacier. Slowly but surely, a mound or dam consisting of boulders, small stones, sand, and clay is eventually formed at its exposed peripheral locations. These mounds of accumulated glacial debris are called moraines. A terminal moraine is formed at the front part of the glacier, and if the glacier has partly melted, the moraine is no longer connected with the glacier and marks the greatest advance of the glacier during the last thousands of years. According to a nearby town and village or a well-known location, geologists have given names to these moraines that are typical for individual glacial periods of the Pleistocene. Today, the majority of all well-known moraines are now on dry land, significantly distant from the glacier. Their distant locations are caused by the fact that during the past 11,500 years, interglacial temperatures were 5–9 °C higher than those in glacial periods and lasted long enough to allow a partial melting of the glacier that originated during 100,000 years of the last glacial period or by the fact that it was existing from the earlier glaciation and this last glaciation continued only within the earlier ice accumulation. Many warm periods even with higher mean temperatures existed during the past hundreds of thousands of years when glaciers, surviving up to now, partly melted and shrank and then grew again in cool periods. Their dynamics of growth and shrinking is traceable by moraines. When a glacier partly melted, the space between the moraine and the receding tail of glacier was entirely adequate to hold the water originating from the melted ice. Hence, many mountain lakes evolved and grew. The number and size of those past lakes prove that the regress of glaciers in some of the earlier interglacials was more distinctive than now. Considering the above, we conclude that the present-day worries of some climatologists about the complete melting of alpine glaciers due to recent warming are groundless.

We have mentioned earlier that the majority of recent soils did not originate directly from weathered rock. They developed in place on unconsolidated sediments that were deposited by floodwaters or on deposits initially weathered at higher elevations and further eroded and moved downhill by water. Even if both processes appear identical, they are not the same because the floodwaters transported the material from great distances and deposited them on large areas, while local erosion was restricted to small distances.

Another alternative was the transport of the weathered material by wind during dust storms occurring in the past. Dust storms were very frequent and extremely strong during as many as 11 major glaciations in the last 2.5 million years. Here, we limit our discussion to the last two glaciations. The earlier one called either Riss in Central Europe (Saalian in northern Europe) or Illinoian in the USA lasted roughly from 200,000 to 130,000 years BP. The next new glaciation denoted as Würm in Europe (Weichselian in northern Europe) or Wisconsin in the USA was closer to recent times, again roughly in the span from 110,000 to 11,500 years BP. These two glacials were separated by the warmer Eemian interglacial (Sangamonian in the USA). Actually, we are now living in the interglacial in the Holocene that started 11,700 years ago and shall end in a near future. Here, referring to a near future, we are speaking in geologic terms where the time scale is at least a thousand times greater than our contemporary scale of human beings. Moreover, our indication of time for glacial periods may change slightly owing to the region.

Because the soil surface was not sufficiently covered by vegetation in the zone south of the great glaciers of the Northern Hemisphere during the glacial periods, it was not protected from blowing winds. Fine soil particles were sucked into the air by hurricane-like winds and deposited hundreds and thousands of kilometers away from the place of their origin. Moreover, with huge volumes of seawater being frozen and kept in glaciers, the sea level sank by about 100–150 m. The muddy sea bottoms emerging along continental shelves exposed to dry, cold atmosphere started to be the probable source of the majority of dust particles in storms. The dust was deposited as loess in thicknesses up to several tens of meters. Today, the largest areas of loess occur in China and have thicknesses greater than 100 m, mantling hillsides and forming extensive loess plateaus. Large loess layers also exist in Central Europe, Ukraine, and a portion of US prairies.

When we mention dust storms, we cannot neglect the mechanical abrasion of rocks and all obstacles protruding above the terrain by sand and dust particles carried by the extremely strong winds in arid regions. An instructive example is the Giza sphinx close to Cairo. Today, the sphinx's features appear hazy and not distinct. Such an appearance was not the intention of the sculptor about 4,500 years ago. He did not do it on purpose. We judge in accordance with excavations and according to statues protected from the outside atmosphere that the sculptor created all details of the sphinx with great precision and clarity. Its contemporary shape is the result of dust storms acting for more than 4,000 years. The sphinx had the same fate as big rocks – bumps and folds of both were smoothed and differed only in the extent of their abrasion. The surfaces of rocks are more smooth since the abrasive action of the dust lasted longer, while in the sphinx we still recognize details of the face. We are not speaking about several injuries or wounds caused by men, soldiers in the last two or three centuries. Since the sphinx was carved out of a sedimentary rock having certain layers softer than others, the softer layers have been more abraded to create horizontal stripes not as a desirable decoration but a simple product of nature.

5.1.2 Chemical Weathering

Chemical weathering causes distinctive changes of rocks and minerals. Principally, it destroys crystal lattices of minerals and dissolves cementing materials of sedimentary rocks that "glue" various mineral particles together. The weathering action on the crystal lattice of a mineral can be demonstrated from an example and experience of our practical life:

The owner of an old house sells it to a developer who decides to increase the strength of its external framework and simultaneously improve its appearance with an attractive façade. All the way around the house, the workers first construct a scaffolding made of tubes that were linked and fixed together with screws and special connections. With the tubular framework in place, the workers carried out the task of annexing a continuous grid to the scaffolding. After adding water to a dry mixture of plaster, they covered the grid and its framework with the moist mixture to form an attractive façade. Having completed their work in a perfect manner, they eagerly awaited the next day to view the dried texture and color of their handicraft. The crystal lattice is represented in our story by the entirety of the constructed scaffolding with its atoms modeled by the special, screwed connections. The scaffolding tubes represent the forces binding the atoms together within the lattice. The dry plaster mixture is an assortment of products from earlier weathering. During the following night, a strong windstorm engulfs the modernized house. Although the initial waves of wind vigorously vibrate the scaffolding, its tubular internal structure cannot endure the repetitive strikes of wind pressure and, as a result, suddenly collapses. This external action of forces imitates in our model the external factors causing weathering. The workers are disappointed the next morning to find a big pile of broken and bent tubes. In our weathering model they are the forces keeping the atoms in the regular form of the crystal lattice. Their absence means that the atoms are liberated from their fate to stay at required positions within the lattice. There are also separated scaffolding connections with embedded screws and even many individual screws scattered in and about the pile. They represent freed atoms no longer attached to the lattice. In addition to these separated parts, there are large, intact remnants of the recently plastered scaffolding in the pile. They demonstrate portions of the mineral not affected by weathering. All of the materials in the pile are mixed with the powder of plaster that originated from the dusty products of earlier weathering.

A more complicated explanation of the silicate crystal structure in soils follows even though we tried to simplify the subject as far as possible. Although less patient readers may jump over these paragraphs and continue to read near the end of this subchapter, they should keep in mind that chemical weathering causes small complete parts of an original crystal to be torn out of the lattice by breaking the weakest bond. At the same time other parts of the crystal lattice are completely disturbed with individual ions being released to form simple solutions in water that are frequently washed away. On the other hand, they could be involved in the creation of a new mineral. What are left in place are amorphous forms, materials without definite configurations that are frequently transformed into new minerals of submicroscopic size, the clay minerals. The great majority of all such fragments and mineral transforms obey the same basic principles – those of the crystal lattices of silicates.

The most frequent minerals in the Earth's crust are silicates and quartz. Because quartz is very resistant to weathering, we have the tendency to say that it does not weather at all. Here, we shall describe weathering of silicates without concealing that other minerals are either more or less easily weathered. But silicates are the majority of all minerals, and when their weathering produces new minerals different from those in rock, we call them secondary minerals. We shall show that they have extreme influence on the physical and chemical properties of soils.

The structure of crystal lattice of minerals is determined by the principle of the most dense arrangement of ions composing the mineral. In order to make the rather complicated structure more transparent and applicable in our popular writing on soils, we shall use a most simplistic approach even though it brings along a danger of pitfalls commonly related to oversimplification. By assuming that the volume of space occupied by ions is spherical and formed by a force field impervious to neighboring ions, we can apply and use a model of solid spheres. The ions – spheres of silicates – are arranged into simple geometric configurations. First, we demonstrate the configurations of centers of ions forming the lattice.

There are pyramidal forms composed of four triangles – one is considered the base and the other three triangles are the walls or faces that share a common vertex. Any of the four faces could be considered the base. Although this form is easily recognized as a triangular pyramid, its scientific name is tetrahedron, derived from the Greek *tetra* meaning four, *hedra* meaning face or seat, or *hedron* meaning having bases or sides (Fig. 5.1).

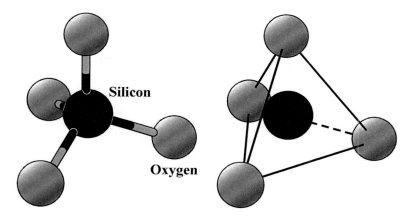

Fig. 5.1 The schematic demonstration of Si tetrahedron in minerals

If one tetrahedron shares its one vertex with that of another tetrahedron, they form twins. If tetrahedrons share two vertices, they form a chain. Three and more tetrahedrons have the capacity to form a plane where each vertex is shared by neighboring tetrahedrons. This characteristic sharing of vertices leads to a great variation of individual lattices. The second type of geometric configuration is the octahedron with eight faces, or walls. Using Greek terminology, we replaced *tetra* by *octa* which means eight. We can obtain this eight-sided configuration when we join the base of a classical form of an Egyptian pyramid to another Egyptian pyramid with its top vertex placed to the opposite side of the vertex of the first pyramid. Of course, the size of these Egyptian pyramids is on the microscale of atoms. Octahedrons have the capacity to share their vertices with neighboring octahedrons or tetrahedrons as we have shown for tetrahedrons. The vertices of geometric configurations are occupied by spheres that are the models of big anions – mainly oxygen O. In some instances, a sphere representing the hydroxyl anion OH occupies the vertices. In the center of the tetrahedron or octahedron, smaller spheres of a cation are located. In the center of the tetrahedron, silicon Si is located in the cavity between the anions, while for the octahedron it is either Al or Fe and Mg.

Silicates are held together by ionic and covalent bonds that occur roughly in equal proportions. Let us now explain the models in more detail. A silicate tetrahedron is formed by a triangle of three O spheres above which is a fourth O sphere seated on the shallow triangular depression of those three O spheres. The cavity below the fourth O sphere is filled by a Si sphere carrying four positive charges, while the O spheres on the four vertices carry a total of eight negative charges. This lack of balance between positive and negative charges accounts for the linkage with neighboring configurations. Individual tetrahedral configurations bind themselves into bow tie, ring, chain, double chain, and sheet silicates; see Figs. 5.2, 5.3, and 5.4. Octahedral configurations are usually seated on the top vertices of a tetrahedral base. With the exception of Na^+ or K^+, a cation such as Al^{3+}, Mg^{3+}, Fe^{3+}, Fe^{2+}, or Ca^{2+} is again in the center of each octahedron (Fig. 5.5). Let us denote them all as the

Fig. 5.2 The arrangement of
Si tetrahedrons in a chain

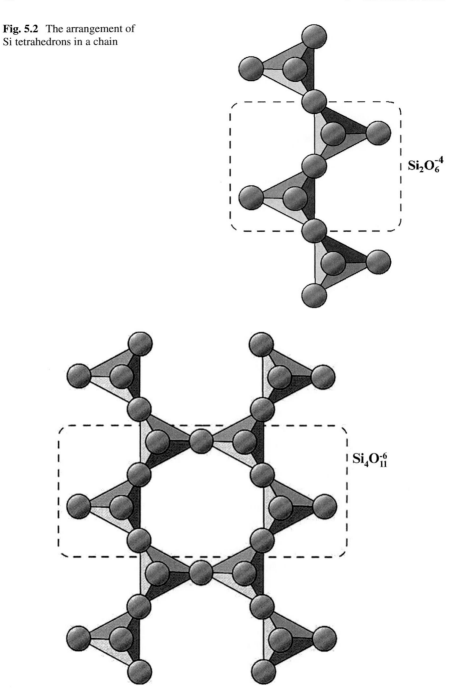

$Si_2O_6^{-4}$

$Si_4O_{11}^{-6}$

Fig. 5.3 The arrangement of Si tetrahedrons in a double chain

Fig. 5.4 The arrangement of
Si tetrahedrons in a plain: A
tetrahedral sheet

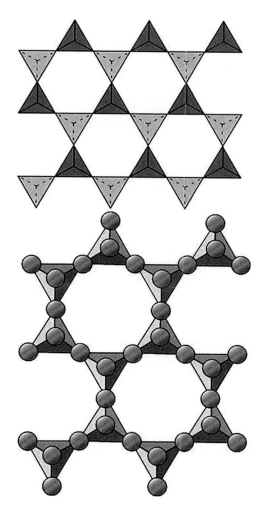

binding bridge B. The vertices of an octahedron are again occupied by spheres of O
or OH whenever they are not coincident with those of the tetrahedron below.
Inasmuch as the bond O–B–O has less energy than that of O–Si–O, its presence
provides a favorable, less resistant location for the initiation of weathering. If we
remember that chemical weathering is mainly the braking of chemical bonds, we
can relate the intensity of weathering to the relative number of weak B bonds. Let
us consider minerals having 24 O atoms linked within each of their structural units
that contain a maximum of 12 B bonds. Accordingly, the degree of lowest to highest
weatherability of these minerals is estimated by a scheme of the number of B bonds
in each of their structural units ranging from 0 to 12. It is usually necessary to
improve this scheme by considering one additional characteristic – the individual
type of cation. For example, the chemical bonding energy between O–Al–O is

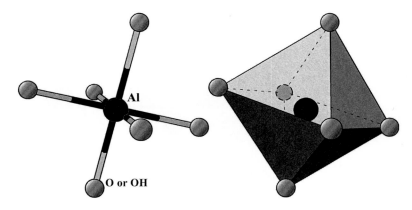

Fig. 5.5 The schematic demonstration of Al octahedron in minerals

higher than that between O–Ca–O. The olivine group with a ring arrangement of silicon tetrahedrals has 10 B bridges in one structural unit. Small differences in its easily weathered characteristic are caused by its various kinds of central ions. Pyroxenes with 8 B bonds have single chains of tetrahedrons linked together with octahedrons. Their weathering is less intensive when compared to minerals of the olivine group. Amphiboles having two chains of tetrahedrons linked together in a unit containing only 6 B bonds are more resistant to weathering than pyroxenes. If the tetrahedrons are linked into a layer, groups of mica minerals are formed with 4 B bonds that are difficult to weather. On the other hand, their weathering proceeds a little more readily if they exist in very thin sheets. Having bonds of only O–Si–O, quartz has the general formula of SiO_2. With octahedrons always missing and in the total absence of any B bonds, each of its oxygen atoms is shared between two tetra-hedrons linked together in all three dimensions. This configuration is why minerals of quartz group are not weathered (Fig. 5.6).

When we return to our example of scaffolding at the start of this chapter, B bonds could be modeled as thinner, even damaged tubes. It is obvious that thin previously damaged tubes will be the first to break when hit by stormy winds, and at the moment they break, they cause a gradual collapse of the whole scaffolding. The higher the number of those broken and unattached tubes, the earlier the collapse occurs. If there were only one or two such tubes, the scaffolding would survive the windstorm without visible damage.

Let us now return to our discussion of crystal lattice weathering. In real minerals the shapely, neat regularity of the described models is disturbed by substitutions of their central cations. For example, a three-valent aluminum cation could take the place of a four-valent silicon cation within a tetrahedron. In spite of their very similar sizes, the stability of such a tetrahedron is disrupted. In octahedrons the central aluminum could be substituted by a two-valent iron or calcium cation accompanied by a loss of stability in a manner similar to the result of substitution in a tetrahedron (Fig. 5.5).

Fig. 5.6 The three-dimensional arrangement of Si tetrahedrons in the crystal lattice of quartz

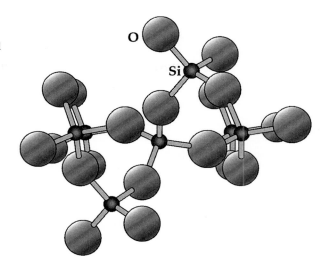

It is a typical feature of nature that total regularity without any substitutions exists only in our models. Under natural conditions of reality, substitutions like those illustrated above weaken the bonds at specific locations within the crystal and satisfy the aims of weathering. The higher the number of those substitutions as well as other irregularities, the easier the mineral is weathered. And the more of those "weakened" minerals appearing in a rock, the easier is the rock's weathering. These substitutions in a crystal lattice, possible only when the exchanging cations are nearly the same size, are called isomorphic substitutions. Such substitutions are responsible for changing the electrical charge on secondary minerals. For example, when the smaller Al^{3+} replaces Si^{4+} in a tetrahedron, the excess of the negative charge appears on the surface of a mineral. And the smaller the mineral, the stronger is the final effect with the net negative charge on the surface of the mineral being more distinct. Similarly, the isomorphic substitution of Al^{3+} by Mg^{+2}, Fe^{2+}, or Ca^{2+} each creates a distinct negative surface charge density.

The role of aluminum in isomorphic substitution is worth of a detailed study in origin and evolution of soils. Aluminum can have two roles in a certain mineral: once it is substituting, and at a just slight distance it could be substituted. These two different roles could be a factor, which makes the processes less uniform and leads to the formation of intrigue plots of the play of mineral transformations in the soil evolution.

Weathering is frequently described as the combination of four basic chemical processes: hydration, hydrolysis, oxidation, and reduction. Although we could add two more, dissolution and precipitation, both of them have to precede together obeying explicit chemical reactions that lead to new simple solutions. Hence, we eventually include them into one of the basic processes – the transformation of the products of weathering.

When the weathering is very strong and when it is realized in acid conditions, the original crystal lattice could be completely disturbed and the "central" cations like Fe^{3+} and Al^{3+} are pulled out to form sesquioxides Fe_2O_3 and Al_2O_3 and eventually transported at least from the place of origin. The transport of sesquioxides is dependent upon several soil conditions as we demonstrate later in Sect. 12.2 about soil horizons.

5.2   The Old and New Minerals

5.2.1   Size of Soil Particles

The study of soil particles, their size, shape, and role upon soil properties belongs to the branch of soil science called soil texture.

Soils are comprised of particles having a huge disparity of sizes. Gravel, the remainder of stones after weathering having individual sizes larger than 2 mm, is also called the skeleton of a soil. We have the opportunity to separate the skeleton from a soil sample by sifting the fine particles through a sieve with circular openings 2 mm in diameter. Observations and careful experiments have shown that gravel composing the skeleton does not support plants – indeed, plants do not even start to grow in particles greater than 2 mm. On the one hand, mixtures of weathered particles less than 2 mm definitely provide some partial support for plants. The size of 2 mm is not an especially accurate value and only represents a reasonable boundary obtained by convention for a life-supporting environment. If gravel prevails in soils, the volume suitable for rooting of plants is strongly reduced, the amount of water being supplied to roots is very small especially during periods without rain, and a feeble import of nutrients usually persists. On the other hand, the presence of even a small, hardly noticeable gravel skeleton can have a positive, highly beneficial role providing relatively big pores along the boundary between its gravel and the surrounding fine soil. Frequently, these big pores are not filled with water even if the bulk of the soil is fully water saturated. In such cases, the big pores form a network through which air can penetrate into the waterlogged soil and provide an opportunity for aerobic processes to exist in the neighborhood of the gravel grains. These large pores provide still another advantage by making it easier for water from especially heavy rains to infiltrate the soil surface. Further on in this chapter, we deal with characteristics of soils without gravel grains, simply with fine soil.

The size distribution of particles in the fine soil is truly remarkable. Extending from as large as 2 mm to as small as less than 0.002 mm (i.e., 2 μm or 2 micrometers), their diameters have a range greater than a thousandfold. The coarse fraction is sand, the finest fraction is clay, and between them is silt. The size of these fractions was originally estimated empirically. For example, after a sample of dry soil was initially weighed, vigorously mixed with an excess of water, and allowed to settle for a couple of hours, the muddy water above the soil sediment was poured out. With the muddy water containing only fine clay particles, it was believed that they would always float and not sink into the sediment. Another excess of clean water was poured into the soil sediment and vigorously mixed, and again after a couple hours, the muddy water was poured out. This mixing and washing process continued until the water above soil sediment was clear and no longer muddy after a couple of hours of settling. After drying the soil sediment, it was weighed, and the difference between the first and second weights was assumed to be the weight of clay in the sampled soil. After hydraulics, the new subject of mechanics, was developed from the end of the eighteenth century to the start of the nineteenth century from Newton's three laws of motion, physicists formulated the velocity of sedimentation of spherical particles in water with a mathematical equation. With sedimentation velocity depending upon the square of a spherical radius, we know that if the radius is 1 and the velocity is 1 as well, then we also know that a sphere of radius 10 has a velocity equal to 100. Soil chemists, who finally specialized in soil physics, applied the equation for determining the size of soil particles.

The first objective classification of soil fractions was elaborated by the Swedish soil chemist Atterberg who assumed even distributions of particles within the following fractions indicated as equivalent sphere diameter: 2–0.6, 0.6–0.2, 0.2–0.06, 0.06–0.02, 0.02–0.006, 0.006–0.002, and <0.002 mm. Having considered their practical utility, he eventually simplified his classification system to *coarse sand* having particle sizes 2.0–0.2 mm, *fine sand* with size ranges of 0.2–0.02 mm, *silt* characterized by the size 0.02–0.002 mm, and *clay* with size <0.002 mm. The boundary between sand and silt varies in national classifications. The US pedologic classifications have set it at either 0.05 mm in the past or 0.02 mm in the recent *Keys to Soil Taxonomy*. In an oversimplification we could say that sand particles are recognizable by naked eye, silt particles are visible by means of lens and optical microscope, and clay particles are examined by an electron microscope. While the general properties of sand and clay fractions are obvious, the silt fraction is characterized as the range from the boundary when sand assumes some claylike properties to the boundary where the main properties of clay come to an end. The latter boundary is demonstrated whenever Brownian movement in water is no longer identified for particles larger than 0.002 mm. The Atterberg system of particle fractions was further modified and slightly differs in the USA from the European system or from that of the International Union of Soil Science. More about the recent classification of soils according to their content of sand, silt, and clay will be discussed in one of the next chapters, in 5.2.3 Soil's First Name: Texture. Because the systems are not everywhere the same and manifest slight differences in various individual countries, we shall simplify them into only five groups before we describe briefly the dominant properties of each of them.

5.2.2   The Finest Minerals

Clay particles have strange properties and characteristics when compared to silt and sand particles. First of all, they do not have a spherical shape or the shape approximately modeled by a sphere. Second, they are not grains. And third, although they could be closely compared to flakes or to very thin disks, each of them has a specific crystal lattice. Indeed, they are *clay minerals*. We authors do understand that it looks like a sort of typing error when two words standing in opposition in our instinctive imagination are put together: *clay* and *minerals*. When we stir wet clay in our fingers, we get the feeling like stirring butter. But minerals have sharp edges, we know it by our observation of many silicate crystals in museum or when we visited the collection of crystals of our friends or when we watched the expensive gems in the store windows of the jewelers. What happened with the sharpness of minerals if the size is decreased to such a minimum as less than 0.002 mm or in other units less than 2 μm? Wet clay is well formed keeping the form even if dried in an oven, and this baking fixes the form, while sand consisting of visible minerals cannot keep the form when it loses only a part of its original water. All those contradictions will be explained rationally when we apply the recent advanced methods of study of minerals.

When I (MK) studied about 55 years ago to learn how clay minerals influence water and as a consequence the water regime in soils, my friends studying literature theory, history, or philosophy at the Charles University used to ask me about my postgraduate studies at the Technical University and what kind of balmy research did I explore. When I responded that I made experiments with clay minerals to understand their influence on the physical properties of water in soil, they were thinking aloud that I bound one clay mineral to a cotton thread, immersed it in water, and observed what was happening. I tried to explain to them that nothing like that was feasible. I attempted to persuade them that even the thinnest thread was extremely thick for a particle of clay and that their imagination was on the level of thinking that I tried to fasten an ant to a hawser or anchor rope. Some nodded their heads in agreement and asked if I used technical equipment like a magnifying glass, while others, not asking at all, whispered that I was "around the bend" or "tetched" in the head.

The knowledge that all clays are not the same was known in China 3,600 years ago. They knew how to select specific places between various localities having whitish clay that had properties of plastic paste after being wetted. The Chinese formed the paste into a cup or some kind of small sculpture that kept its original shape after being dried and baked. Improving these procedures, they produced the first porcelain about 2,200 years ago. The technology continued for 1,600 years until the Chinese porcelain was imported to Venice, to the Medici family, and to the courts of kings in Europe. People were shocked by the intricate porcelain shapes and by the near transparency of the walls of cups. Astonishment and wonderment of the Chinese products was so strong in England that the nobility referred to the newly imported goods as *china* and the term started to be broadly used in English. The raw material used for the production of porcelain is clay containing predominantly the mineral kaolinite. This Chinese clay was originally mined in the province Jiangxi in the hilly country Jingdezhen. The name of the locality is Gaoling, translated as "high hill." The mined clay, named kaolin by the French distortion of the Chinese name, became the source of the scientific name of the mineral. One more excursion to linguistics: The name porcelain is derived from the old Italian *porcellana* meaning shell, because of its resemblance to the translucent surface of a shell.

Kaolinites are members of a group of minerals with a lattice basically identical to the form of the clay mineral *kaolinite*. With their main properties being very nearly the same, we shall only describe those of kaolinite that has a crystal lattice typical of silicates. It is a two-sheet mineral with one sheet being formed by silica tetrahedrals overlain by a sheet of aluminum octahedrals. They appear drunk as they lay on their sides. This configuration is denoted as a 1:1 lattice, meaning one tetrahedral and one octahedral sheet are fixed firmly together. Here we denote this 1:1 arrangement as a double layer or briefly as T-O configuration. Its chemical notation is $Al_2Si_2O_5(OH)_4$. Both sheets are solidly and permanently bonded together by sharing mutually common oxygen atoms that are on the top of tetrahedrals and on the corner of octahedrals. The tetrahedral bases are linked together on a plane to form a repetitious hexagonal motif where all corners of one triangle are common with the neighboring triangles. Many of these hexagonal motifs form a large sheet where each central silicon placed into the small cavity formed between four oxygens (in three-dimensional space) balances its three positive charges with negative charges of three neighboring oxygens on the base; and because each oxygen on the sheet has two neighboring silicons, its two negative charges are also balanced. With the fourth positive charge of silicon being balanced by the negative charge of oxygen positioned perpendicularly on the top of the tetrahedron, all charges in the tetrahedral sheet are balanced. The octahedral sheets, positioned on the top of the tetrahedral sheets, share their corner oxygens with the tetrahedral sheets. Aluminum cations of the octahedron have three positive charges, partly balanced with the negative charge of oxygens of the tetrahedron and partly with the negative charge of hydroxyls (OH) on the top corner of the octahedron. Again, the charges are balanced. At a distance of 0.29 nm (nanometers=0.00000029 mm), another double layer 1:1 formation starts again with tetrahedrons (Fig. 5.7).

The double layers are fixed by a hydrogen bond originating in hydroxyls of octahedrals in the neighboring double layer formation. This hydrogen bond can be compared to a married man who keeps his marriage but also is disloyal by having an ongoing relation with another woman. He spends more of his energy at home but enjoys about one-third of it with his girlfriend. Thus, the 1:1 double layer formations or T-O configurations are firmly fixed together by hydrogen bonds. The complete mineral is therefore comprised of many hydrogen-bonded T-O configurations, all of which are also bound to their neighbors by shared hydrogen bonds. The entire kaolinitic mineral can be observed on the screen of an electron microscope. When examined on the screen of an electron microscope, sharp contours of its hexagonal platy crystals, usually smaller than 1 μm, are vividly observed because many of its

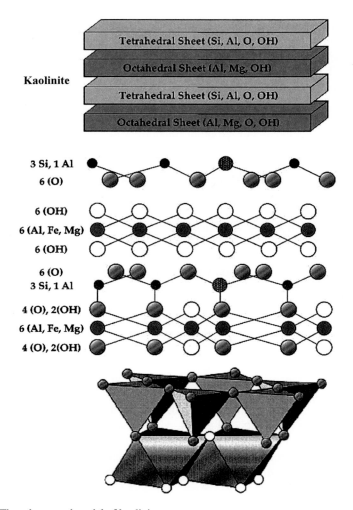

Fig. 5.7 The scheme and model of kaolinite

double layers residing exactly on top of each other are sufficiently thick to render them impermeable to electron radiation.

Kaolinite does not swell when wetted by an excess of water because water molecules cannot penetrate between its double layers and are only adsorbed on its outside crystal surfaces. Therefore, whenever moist kaolinite is dried or even baked in an oven, it never shrinks. Owing to this property, kaolin – the clay containing kaolinite – is used in the porcelain industry. The stability of kaolinite is also attributed to the fact that ions cannot enter between its 1:1 double layers. Its stability is even further guaranteed by its nearly neutral electrical surface charge density attributed to the very small number of free negative charges that can only exist when bonds are broken as a result of its mineral surface being disturbed. This tiny magnitude of free negative charges allows only a very small number of free cations to be attracted to its surface from an external solution of simple chemical compounds.

Kaolinites originated as thick layers in the past during more than tens of millions of years ago due to a strong weathering of minerals in a hot and wet tropical climate. Nowadays, they are continuing to slowly develop in soils of the tropics.

We can compare the crystal lattice of kaolinite to a cluster of slats. One slat is white and it represents the tetrahedral sheet. The black slat representing the octahedral sheet is nailed to the white slat. The nails represent here the oxygens on the top of the T-sheet and the same oxygen belonging to the O-sheet. The thickness of this double layer slats is about 0.4 nm (0.4 nanometers=0.0000004 mm). The free surface of the black slat is covered by the glue that sticks to another white slack that is nailed to another black slat. This arrangement of slats is repeating many times in the same sequence and the glue represents the hydrogen bond. Nothing can penetrate between the mutually glued white-black double layers. Inasmuch as nothing can extend one white-black combination of slats (double layer) from the next combination of slats (double layer), kaolinites do not swell or shrink because water molecules cannot enter the space between double layers. Water molecules are bigger than the space occupied by hydrogen bonds.

Montmorillonite, having many characteristics quite opposite to those of kaolinite, belongs to the group *smectites* and was named after the Canton de Montmorillon in France where it was discovered in 1847. Clays containing predominantly montmorillonite – or generally smectites – swell when they are wetted and subsequently shrink when they are dried. The shrinkage is so intensive that a network of big cracks appears on the surface of a field with soils formed by those smectite clays. The cracks are often deep and large enough to stick our entire hand into them. A detailed description of the mineral montmorillonite isolated from this kind of clay was performed in the third decade of the last (twentieth) century when adequate instruments were available to make accurate measurements by X-ray diffraction and electron microscopy. The first instrument works on the principle that each layer formation behaves like a mirror. Hence, X-rays reflected by the layer surfaces manifested in bands of different frequencies show distances that are specific for a particular type of clay mineral. Electron transmission microscopes are built on principles similar to those of ordinary microscopes but transmit very short waves of electron radiation rather than the long waves of light visible to the human eye. Because electrons have wavelengths about 100,000 times shorter than those of visible light, the resolution of an electron microscope is about one million times greater than that of an ordinary microscope. From today's recently advanced, more sophisticated instruments, such resolution has been remarkably increased by several orders of magnitude. Instead of glass lenses to concentrate and focus waves of radiation, a virtual lens is mathematically manipulated to focus electromagnetic rings of radiation. An object placed against the radiation forms a shadow like we form a shadow during a sunny day. In an ordinary microscope the shadow is observed down to the size of about 0.01 mm. In an electron microscope the shadow is observed down to the size of the object – a clay mineral between roughly 0.005 mm (5 μm) and 1 nm

(0.001 μm) or even less. These techniques supplemented by chemical and thermal analyses provide quantitative information about the structure of a crystal lattice (Fig. 5.8).

As mentioned above, the crystal lattice of montmorillonites was described first. Only later on was the whole family of minerals similar to montmorillonite discovered. They were named smectites derived from the Greek *smectos* which means blurred or painted, because their image in electron microscopy lacked sharp contours and because the clay was smoothly and easily spread between the fingers.

Smectite crystal lattices consist of three sheets: tetrahedral-octahedral-tetrahedral, or briefly T-O-T configuration. Let us call this formation a triple layer. We can say that the octahedral sheet is closed from the top as well as from the bottom side by a tetrahedral sheet. Because of this type of imprisonment, the octahedral sheet has no chance to demonstrate the majority of its own characteristic features. At first it looks as if we merely added a tetrahedral sheet to the top corners of the octahedrals of kaolinite. However, it is not so simple as it initially appears since there is a significant change "in the guts" of smectites, and especially inside montmorillonite. When we go back to slats in our oversimplified model of kaolinite to describe smectites, we have one black slat to which the white slacks are nailed, one at the bottom, one on the top. And although this is the form of the triple layer, there is no glue on the outer side of triple layer. With the glue being absent, the distance between neighboring triple layers in smectites is bigger than the distance between double layers of kaolinite. Glue represents hydrogen bonds in our visual aid of kaolinites. In smectites with no hydrogen bonds between its triple layers, water molecules as well as ions of solutions can enter the space between triple layers that can mutually move and slip a little bit. The more water we add to smectite clay, the more the distance between its triple layers grows, and as a result, we observe a swelling of the clay. When the clay is slowly drying, water molecules disappear from the space between triple layers and the sheet of water molecules becomes thinner. We observe the change as the clay shrinks. Both processes result in minor slipping of the triple layers, one on the top of the lower one. Hence, the image of smectites and especially that of montmorillonites in an electron microscope looks hazy without distinct boundaries. We have to explain now why it is so and why and how smectites differ so much from kaolinites.

In octahedral sheets of smectites, the three-valent aluminum is in some instances substituted permanently by the two-valent cation Fe^{2+} or Mg^{2+}. With their sizes also not being the same, the configuration is rendered into a stage of slight instability that allows the unbalanced negative charge of the sheet to appear on the surface of the triple layer T-O-T. In a similar way the Si^{4+} in the center of some tetrahedral sheets is substituted by three-valent cations like Al^{3+} and less frequently by other lower valent cations. Both types of permanent substitution result in the appearance of the negative charge on the outside plane of each triple layer. The result is not only the absence of hydrogen bonds or their analogy but also an opposite effect. When one plane of a triple layer T-O-T with an excess of negative charge faces another plane of T-O-T configuration with an excess of negative charge, the two triple layers are

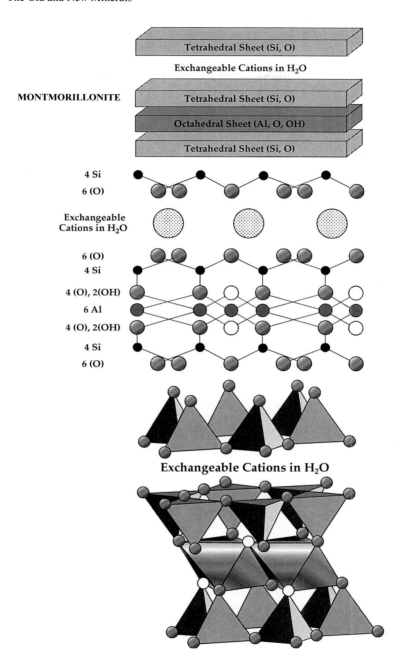

Fig. 5.8 The scheme and model of montmorillonite

mutually repulsed. The same repulsion happens with all such triple layers, i.e., between T-O-T configurations. Provided that there is enough water available, water molecules can simply penetrate into the space between the repulsed triple layers and eventually form a film with a thickness of several water molecules. However, there are still outer edge to outer edge bonds of sufficient force to hold bunches of about 10–30 triple layers together. At this point, we have explained that all volume changes of smectites during their wetting (swelling) and during their drying (shrinkage) are the direct result of changes of water film thickness between T-O-T configurations. Nevertheless, an interesting effect arises during drying after crossing a certain boundary of water content reduction. At that diminishing threshold of water content, the cohesion between triple layers is broken to such an extent that we are finally able to observe with naked eye the appearance of fissures in a clay predominated by smectites.

As water enters into the spaces between triple layers T-O-T, dissolved inorganic compounds are introduced into this negatively charged domain. Cations of the inorganic compounds, attracted to the field of negative charges, behave as if they are weakly bonded to the clay mineral. However, when the solution outside the mineral is replaced by another solution containing another type of cation, the cations from the former solution are replaced by those in the latter solution. Since this replacement or exchange of cations occurs in the space between triple layers, they are called exchangeable cations. Inasmuch as each clay mineral has a certain capacity to retain exchangeable cations, this feature is denoted as cation exchangeable capacity with the frequent abbreviation CEC. This feature or property is used as a simplified measure of soil fertility because it offers a numerical value of the retention of possible plant nutrients commonly expressed as milliequivalent of hydrogen per 100 g of dry soil (meq/100 g). The term milliequivalent describes the weight of a substance in milligrams divided by its valence. CEC is closely correlated to the specific surface (m^2/g). In montmorillonite the external specific surface makes up about 20 % of the total surface, and 80 % of it belongs to the internal surface, i.e., the surface of T-O-T layers inside of the crystal of clay mineral. The total specific surface accessible to water molecules is about 300–400 m^2/g and CEC is in ranges 80–140 meq/100 g. Just for a rough imagination, if you have a handful of clay soil that is 50 % montmorillonite, you have in your hand soil particles that have an internal surface area ranging between about 1,000 and 3,000 m^2 or approximately the size of a family house garden! Kaolinite belongs to clay minerals with the smallest value of the specific surface between about 5 and 15 m^2/g, and its CEC is 3–15 meq/100 g. The values are again very approximate – they depend on the degree of perfectness of crystal lattice and on the environmental conditions when the clay mineral was formed.

Illite is a clay mineral with a crystal lattice also configured by three sheets T-O-T. When compared to smectites, illite's main difference is in the bonding of many of individual triple layers together by the K^+ cation that fits perfectly between the layers to balance the negative charge of the planar surfaces of triple layers that we mentioned when we described the crystal lattice of montmorillonite. We consider the K^+ in this position as fixed and an integral part of its crystal lattice. If all such positions were filled by K^+, the crystal lattice would be close to mica, i.e., to muscovite and biotite. Compared to mica, all of these positions in illite are not filled by K^+ – some positions are free and available to exchangeable cations from the outside solution of mineral salts. The specific surface accessible to water molecules and the value of CEC of illite are both smaller than they are in smectites. In illite, although cations K^+ form an inseparable part of the crystal lattice, they do not occupy all potentially fitting positions between triple layers as they do in mica. Clay minerals in the illite group differ from mica in still another manner. There are more frequent substitutions of central Si^{4+} by Al^{3+} in their tetrahedral sheet as well as more frequent substitutions of Al^{3+} by Fe^{2+} in their octahedral sheets. Therefore, the T-O-T configurations contain an internal unbalanced negative charge that is balanced by exchangeable cations residing on the external sides of the tetrahedrons. As a result, illites have a medium value of CEC. As a permanent part of the crystal lattice, some K^+ cations serve as an obstacle against water molecules entering the entire interlayer space. They must enter only at places where K^+ is missing. Thus, these K^+ cations curtail and restrict entrances of water molecules and exchangeable cations from the outside solution. But this behavior caused by the mutual bonding of triple layers is not always realized. Whenever K^+ does not occupy these potential positions as part of the lattice, the physical behavior of illites begins to approach that of smectites (Fig. 5.9).

It is of interest to also know that the K^+ ion in the lattice of some illites is occasionally substituted by Mg^{2+}. Owing to the difference in size of these cations, the crystal lattice is sufficiently altered to make it easier for both water molecules and exchangeable cations to enter into the triple interlayer space. Although illites slightly swell and shrink, such processes are rather restricted. Their internal specific surface is below 50 % of their total specific surface that ranges from about 40 to 90 m^2/g with CEC values being roughly 10–40 meq/100 g.

When we demonstrate the basic physical characteristics using our slats model, we deal again with a black slat representing the octahedral sheet to which are nailed on both of its sides white slats representing tetrahedral sheets. Up to this comparison the model is identical to that of smectites. But now a principle difference occurs. A small ring representing cation K^+ nails both triple layers together on some positions and balances at least partly the negative charge of each white slat.

The origin of many clay minerals is demonstrated in the simplest way when the weathering of mica is studied. The crystal lattice of mica, formed by triple layers T-O-T bound together by K^+ cations occupying all available positions between two neighboring tetrahedral sheets, is written in an abbreviated form as repetitive triple layers T-O-T-K^+–T-O-T-K^+–T-O-T-K^+ and so on, where the symbol K^+ represents the cation linking the neighboring triple layers T-O-T together. One of the important steps in weathering is the extraction of some K^+ cations from their fixed positions in the interlayers between T-O-T configurations. In this way illites are formed. This extraction is accompanied sometimes with the substitution of the central cation in the tetrahedrals or octahedrals. The more intensive the weathering, the more K^+ are pulled out until all K^+ disappear from their originally fixed positions in the interlayer. In this way smectites are formed. When very strong weathering lasts for a

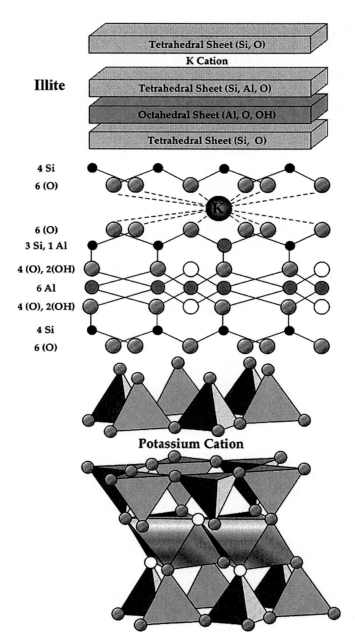

Fig. 5.9 The scheme and model of illite. Potassium is not present regularly

long time, a tetrahedral sheet is eventually broken and displaced without keeping its original configuration. Kaolinites are the result of such weathering. However, the weathering process does not end. It could continue until bauxite is formed with an aluminum cation remaining in the center of its octahedral. Having oversimplified the formation of clay minerals by the process of weathering, we omitted numerous steps of their synthesis that produce a diversity of qualities and properties. These features hinge on the chemical composition of water filling the pores, conditions and frequencies of water drainage or water logging, prevailing acidity and temperatures, and a multitude of other timely and impinging conditions.

Just one group of clay minerals deviates from slats models. They are *allophanes*. They originate as products of weathering of volcanic materials, mainly of volcanic glass. Their name was derived from Greek *allos* meaning other or different and *phanos* meaning to appear. They were earlier classified as mainly amorphous hydrous aluminosilicates, but now they are regarded as minerals formed by fine spheres of 3.5–5 nm outside diameter. These nano-spheres remain empty and have walls formed by tetrahedrals Si–O–OH bound to octahedrals Al–O–OH where the Al octahedrals may prevail on some parts of their spherical structure. The chemical compositions of various kinds of allophanes fall in a relatively narrow range and have SiO_2:Al_2O_3 ratios mostly between 1.0 and 2.0. In addition to their dominant spherical form, some of them also have ring-shaped forms. The specific characteristics of allophanes can be attributed to their spherical shape. Their apparent specific surface is high. They are characterized by high phosphate retention. Their low bulk density is attributed to their empty spherical volume and to spheres having roughly the same diameter. Soils with clay fractions dominated by allophanes have similar physical properties such as a high cation exchange capacity, a high porosity, and a large hydraulic conductivity.

5.2.3   Soil's First Name: Texture

It has been a well-known fact to *Homo sapiens* for at least the last 12,000 years that different kinds of soils have various properties. Their knowledge on soil qualities is linked to the start of the greatest revolution in mankind's history, i.e., to the start of agriculture at the end of Paleolithic period. Actually, this was the time when the Neolithic period began. Without knowledge of soils, establishing a life related to regular agriculture was impossible. Although no documents were written during this era because at that time script had not yet been invented, the style of life, the selection of land for very simple settlements, and the regular planting of first domesticated barley and wheat were all inconceivable without sorting and optimizing land parcels of a large area occupied by the families of settlers. That large area was the region of the Middle East extending from the Nile valley to the Tigris and Euphrates rivers including the slopes of the Zagros and Anti-Taurus mountains where the first agronomists found the best soils as deposits of mountain slopes and fertile alluvial soils in river valleys. Here, archeological excavations have uncovered the first remains of agriculture.

The first written notes by ancient Egyptians were about two types of soil: *ta kemet* and *deshret*. *Ta kemet* was the dark or black soil in the Nile valley that originated and continually kept very fertile by regular muddy floods of the Nile River. But *ta kemet* was also the ancient Egyptian name of Egypt. At that time, it was beyond the imagination of those Egyptians that Egypt could exist outside the alluvial soil of the Nile valley. Behind the boundary of the Nile valley, there was the red desert land or the seat of death, *deshret*, according to Egyptians. *Deshret* was also the formal name for the Red Crown of Lower Egypt. However, the written documentation of a real soil was actually mentioned much later in classical Greek literature and discussed much more in ancient Roman literature. Their dominant attention was toward less fertile soils, namely, to sand which was considered a soil of very low quality. When an ancient Roman intended to speak about needless work and activity, he would say: *Ex harena funiculum non est nectere.* You have no chance to knit a rope of sand. A script based on flimsy, poor arguments was described: *In arena aedificas.* You are constructing or building on sand. It is often found in old chronicles: When one of two sons inherited lands with sandy soils and his brother received through inheritance lands with loamy soils, cruel arguments started and frequently ended in fratricide. Sand was generally considered a poor soil, while loam was regarded as a fertile soil offering high yields. The only one exception was salinization of soils. Ancient Sumerians made substantial distinction between black loamy soil and white soil, i.e., loamy or heavy loamy soil with a sprinkling of white dust having a salty taste. The salt sprinkling was sometimes merging into a thin layer. Such soils offered no yields and were always abandoned (more about them in Chap. 13.1).

As found in the Code of Hammurabi written about 3,800 years ago, taxes were mandated by rulers initially according to yields of natural products estimated from specific ratios of harvest. Tax collectors belonged to high society, but some of them were crooked men increasing taxes and stealing this excess for their own benefit. The best-known tax-collecting crook was Tarim Dagan from the Babylonian town Mari who sought 50 % of harvest that included newborn lambs. He entered history because of complaints of rich noblemen as well as those of poor farmers. Their complaints were written in cuneiform documents on clay tablets. Answers and responses from the Babylonian king to these complaints also exist, but not one word is mentioned about land or soil taxes. Similar situations occurred in later history when monarchs and noblemen taxed farmers in accordance to the extent of lands they used. Moreover, the height of the tax was dependent upon the distance from the ruler's seat or upon various patterns of land usage. Remarkably, the quality of soil in such locations was not considered, in spite of the fact that the farmer knew where he could expect highest yields.

The ancient Greek philosophers were the first to deal with the idea that soil is part of nature, and since nature has many shapes, soils cannot be uniform. Xenophon, two and a half thousand years ago, wrote that life started in soil and is ending there as well. Hesiodos distinguished between various soils according to the type of used plow. In spite of Greek philosophers' intentions to devote their studies on soils to the abstract level, they often compared a soil to a woman giving birth to a new life. Theophrastus (about 2,300 years ago) proposed the first classification of soils according to fragments of his scripts on plants.

The practical Romans took over and extended the Greek knowledge on agriculture related to observation of soil properties. In the second century BC, Cato recommended use of animal manure and green manure to improve soil fertility. Marcus Terentius Varro, a close friend of Cicero, described and classified various soils in his *Rerum Rusticarum Libri Tres* and warned his contemporaries to avoid swamps and marshland that serve as sources of disease. To his merit we have information on the Greek Theophrastus.

With the decline and eventual end of the Roman Empire, all earlier scripts on soil were lost and forgotten until the arrival of Renaissance. However, this comeback was not so easy and obvious. It all started with the new role of money in the society where the first economists did not call themselves so, but who recognized that soil is a tool in production of foodstuffs and that this tool should be taxed instead of harvests.

The first soil taxation attempts appeared at and after the end of the 30-year war in Europe, when the great majority of population was very impoverished, including the lower noblemen and the landlords who distributed lands to themselves if their rulers were winning battles. In the first rustic law, soils were divided into three "bonity" (meaning quality) classes: good, medium, and bad bonity. The developing tax system required a fundamental criterion based on objective details. Using modern terminology, that criterion was "soil class" according to percentages of sand, silt, and clay particles in each soil. Today, we also speak about it as soil texture. As methods for determining soil texture advanced, the more objective was the characterization and class of a soil. We should not forget that it was the tax system that required an objective soil classification system in order that individual governments or countries could have or expand their own system of taxation that imposed taxes on land or soil.

We find now the following terms in the US system: sand, loamy sand, sandy loam, sandy clay loam, sandy clay, loam, silt loam, silt, silty clay loam, clay loam, silty clay, and clay. The classification depends upon the content of three groups of particles characterized by their sizes: sand (2–0.05 mm) and silt (0.05–0.002 mm or 50–2 μm). In some classification systems, the silt is in ranges 0.02–0.002 mm. Then there the sand reaches up to 0.02 mm. The size of clay is in all systems below 0.002 mm, or < 2 μm. If there are particles bigger than 2 mm, they are denoted as gravel or skeleton and their amount is expressed in %, but the amount of sand, silt, and clay is determined in the laboratory after gravel particles are separated by sieving. Therefore, the percentage of sand, silt, and clay is measured and calculated as if there was no gravel and as if particle shape is nearly spherical (Fig. 5.10).

The group of sandy soil, loamy sand, and sandy clay loam is sometimes called (mainly in verbal communication) as light soils, while clays and soils near to clays are denoted as heavy soils. Do the scientists as well as laymen wrongly refer to "light" versus "heavy" soils when they are talking about coarse-textured versus

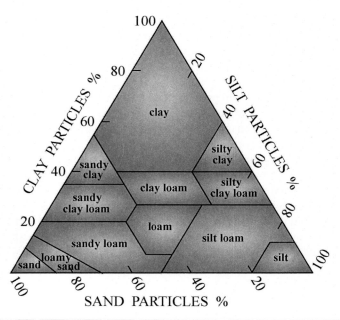

Fig. 5.10 Textural soil classification triangle used by soil scientists in the USA

fine-textured soils or vice versa? It is not a big point, but we know that dry sandy soils are usually heavy not light and clayey soils, if dry, are light not heavy compared to sandy soils. So as a result when someone talks about a heavy clay loam, it is ambiguous. He may mean a loam with lots of sand or a loam with lots of clay. On the other hand, because it takes more force to plow, break up, and turn over the topsoil of a clay than do the same with a sand, clay soils are usually called heavy soils and sandy soils are usually called light soils. Soils between light and heavy soils are then named medium soils.

The textural classification of soils was the first attempt of soil scientists to develop a useful jargon for identifying and naming soils on a regional basis. The jargon became more objective as institutional cooperation in soil science was gradually established around the world. However, soil texture is nothing more than just the first prerequisite for further soil evolution. Since we like to use metaphors, soil texture can be compared with the genes of animals. In other words, although soil texture shows the limits of how and where a certain soil can be further developed, the ultimate or slowly ever-changing characteristics and properties of a soil depend upon the action of many other factors existing in the local climatic environment. This was why we used the title "Soil's First Name" of this short chapter. The real

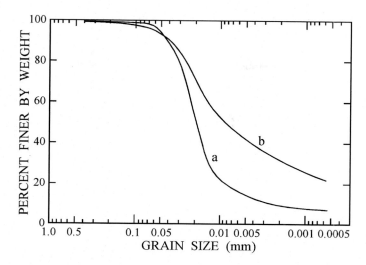

Fig. 5.11 Particle size distribution curve: (*a*) is silt; (*b*) is silty clay loam

and more complex soil taxonomic determination will be explained in the second half of our book.

Because the verbal designation of soil texture is not yet unified in all countries, it is advantageous to plot the results of our measurements in a graph where the size of particles is plotted in logarithmic scale on the horizontal axis and the percentage of particles on the vertical axis. The curve is a summation curve, usually referred to as cumulative particle size distribution curve. We explain the principle of reading and explaining this curve by the following example. For a particle size of 0.02 mm, the curve passes through a cumulative distribution value of 85 %, and for a particle size of 0.002 mm, the curve passes through a cumulative distribution value of 40 %. Thus, the silt content is (85–40)=45 %. The soil texture is close to that of a loam. The more numerous are the groups of particles defined according to their size, i.e., the narrower is the size of particles in the group, the more smooth is the curve plotted in the graph (Fig. 5.11).

The principle of measuring the percentage of individual soil particle fractions is based upon the sedimentation velocity of individual particles in stagnant water. Following the most frequently used procedure, we carefully stir soil in water, use a chemical procedure to break up every aggregate into individually separated particles, pour this suspension into a calibrated cylinder, add water, mix it well, and measure the velocities of particles by inserting a specially designed densitometer – a hydrometer – into the suspension. The density of the suspension decreases with time as particles of different sizes individually settle at different rates toward the bottom of the cylinder. During early stages of measurement, the density decreases quickly because the large sand particles settle quickly. Silt particles, because of their smaller size, settle more slowly and are kept in suspension for longer times. Clay particles, the smallest, move very slowly and even after a day remain in suspension without causing an observable density change in the suspension. In addition to the classical hydrometer, more sophisticated methods are used, e.g., measuring the density of suspension by X- or γ-ray attenuation and laser light scattering or by using laser light to count the individual particles.

5.3   Why the Topsoil Is Black or Gray Brown

5.3.1   Soil Humus

The birth of soils is not thinkable without the existence of living bodies. Up to now we devoted a lot of time to the physical and chemical transformation of the mineral part of the thin top skin of our planet Earth while temporarily paying no attention to the existence of plant life. Its existence is closely linked to soils, and real soils are full of living organisms nourished either directly or indirectly by plants and their decaying and decayed remnants. Knowing how to repay these living soil organisms, soils graciously support all forms of higher plant life. Moreover, the destiny of dead plants or their parts is intimately linked to numerous, diverse processes occurring within soils and to the existence of soils.

The birth of soils continued as the earliest of plants were caught at the initial appearance of fine cracks in solid rocks on the surface of the Earth – indeed, their thin and fragile roots had just gained a major victory against the massive rocks. As these nets of cracks or fissures increased, plant numbers and vegetative mass increased. Their seeds and dead bodies contributed to the extension of growing plants. If the seeds fell on a real soil and not merely into a crack, they had an easier responsibility to transform themselves. On the other hand, the dead, decaying plant parts had in principle a closely related future: their own transformation that we call *humification*. Due to their transformation, the top horizon of a soil often has a dark color in spite of the relatively low concentration of organic matter found in the great majority of soils. Even in fertile soils, the amount of decayed and humified remnants of plants is an order of magnitude smaller than the amount of mineral matter. Speaking about percentages, the total content of organic matter in the top horizon of many soils is about 3 % by weight and ranges from less than 1 % in desert soils to about 5 % and in prairie soils. If it is higher but still below 50 %, we speak about peaty soils and adequate terms according to the majority of taxonomic systems, and if the content is above 50 %, then the soil is called peat.

The transformed organic matter in soil is humus. The term is derived from the Latin *humus* meaning earth or ground. Humus is also a gourmet spread that originated in the Middle East (in Arabic *hoummous* meaning chickpea) and has a variety of spellings. In popular speech of many languages, humus is used in still another completely different sense; it means a great mess. During my (MK) travels over the world, I once had an accommodation in a bungalow for a couple of months. As my Yugoslav colleague who was a mathematician, not a soil scientist, pointed his forefinger to the filthy stove smeared by decades of unwashed fat, he exclaimed: you mean to stay here in this humus. It was of interest to me that the principle linguistic root of humus was kept in his exclamation: a significant proportion of the mass of the stove was formed by organic matter. On the other hand, had it been soil humus, I would not have had to tolerate its stench that filled every room in the bungalow.

That soil humus does not stink is one of its many positive features. The experienced farmer recognizes the quality of his soil after he wets and stirs the soil between his fingers and merely sniffs its aroma. Soil humus improves soils rather than belonging to materials that depreciate them. With the degree of soil improvement depending on the characteristics of humus, farmers and soil scientists maintain their sniffing ability to diagnose aromas into an array of humus qualities ranging from mediocre to top notch. Soils rich in high-quality humus are very fertile and have many properties useful for plants and for many factors of the environment that include hydrology, landscape stability, and air quality. No matter where a person lives within a local environment of any continent – on a farm, in a small village, or in a huge metropolitan region far from a farm – the aroma from a moist fertile soil never fails to stimulate the pleasant feelings and thoughts of life.

5.3.2   Humification

As we mentioned already above, the transformation of all organic matter into humus is called humification. Sources for this process are the remnants of plants originally formed by photosynthesis like leaves, needles, plant stems, branches, and trunks of trees, everything that grew above the surface and fell on the surface after dying off. Added to them are organic wastes and sewage from human activities. From the broad beginning of agriculture about 8,000–5,000 years ago, manuring soils with dung or composted waste was a very important practice for improving and sustaining soil fertility. Ancient Greek and Roman authors described the importance of composts. For example, the essay *De Agri Cultura* (About Agriculture) was a farmer's manual for good day-to-day management and purification of a farm written by M. Porcius Cato (160 BC). But it was more than hundred years later when Marcus Terentius Varro discussed much greater details including the procedure on how to reach a good-quality compost in his *Rerum Rusticarum Libri Tres* (Agriculture Topics in Three Books).

The great majority of all plant remnants lying dead on the soil surface were drawn into the soil during humification with virtually none of them left on the surface. In addition to these "top" materials, organic bits and pieces consisting of decayed plant and tree roots, dead fungi growing originally on the roots, dead microorganisms, and dead soil fauna were all involved in the humification process occurring below the soil surface. Groups of living micro- and macroorganisms attacked all of those organic substances with each group specialized to decompose certain types of waste characterized by their composition of peculiar organic compounds. For example, fungi are the main group decomposing wood that is always rich in lignin, while the remains of cultivated plants with lots of cellulose are

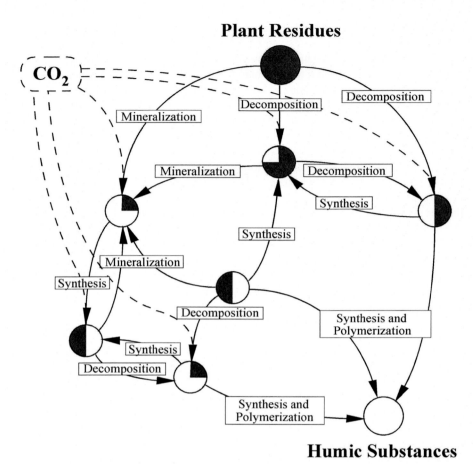

Fig. 5.12 The scheme of the humification process in the soil. The release of CO_2 accompanies decomposition and mineralization. It is presented as the *broken line*

degraded by bacteria. The products of decomposition are accompanied by newly created organic molecules that appear in organic films covering soil particles, which are predominantly clay minerals. These new organo-mineral components of soil change the physics and physical chemistry of soils (Fig. 5.12).

Earthworms, digesting these and other organic nutrients together with clay minerals, contribute to the continual mixing of humification products. They, together with ants, are the primary actors drawing crude remnants of plants from the surface down into the topsoil.

Individual components of organic matter are not decomposed and transformed at the same rate. Organic acids, starch, glucose, lactose, and fructose are decomposed quickly, while proteins and fats are decomposed slowly. Lignin, wax, and resin are very resistant. Some organic compounds are partially incorporated into soil together with dead remnants, while others are by-products of humification. As a result, in the humic top horizon of soils, we find vitamins, enzymes, and even some antibiotics that influence in many ways not only living soil organisms but also plant roots and growth rates of plants. These multiple influences are usually positive and only in some instances cause a negative effect. Phenols and a variety of organic acids appearing mainly in the acid medium of waterlogged soils cause the growth of plants to slow down. Their action also slows down the transformation of plant remnants and acts unfavorably upon microbial life. Fortunately and quite the opposite, that sort of humification also leads to the origin of humic acids that act positively to increase the permeability of cell membranes and therefore enhances the acceptance of nitrogen and other important nutritional elements by the plant root system.

Humification is not limited to merely the decomposition of organic matter. Once it starts, large organic molecules are transformed and cut into smaller molecules. One portion of them is again synthesized into more complex organic molecules under the contributing action of microorganisms. The remaining portion does not enter into any new synthesis, but continues to be decomposed. The portion of newly synthesized molecules also accompanies this decomposition. In other words, the continuation of decomposition is realized in new different ways than it was during the initial process. Some processes appear to run in a circle. Simpler organic compounds are created due to the decomposition of very complex organic materials. Parts of these simpler compounds enter a new synthesis resulting in formation of a group of various complex compounds that have substantially different properties than the "mother" compounds. The processes of decomposition and synthesis are repeated, and with their repetition, all of the properties of the newly formed compounds differ more and more from the initial organic matter. With the origin of new compounds not being restricted just to synthesis, other concomitant processes formed chains that are coiled into a ball at the molecular scale. All such processes are accompanied by the release of simple inorganic compounds that are dissolved in soil water. And if any of these solutes were plant nutrients, they were immediately doled out for absorption by plant roots. Hence, groups of humic substances are formed as a mixture of many molecules, many of which have an aromatic nucleus with phenolic, quinone, and carboxyl components linked together.

We have demonstrated in a very simplified way that the products of humification, mixtures of stable polymolecular material, are sufficiently resistant against further decomposition to prevent their return to cycles of decomposition and resynthesis. Being out of the reach of these processes, they form a new family of soil material even though it is not completely saved from further decomposition. Their continued decomposition rate is an order of magnitude slower than the decomposition rate of the original raw organic material. In order to more clearly express the decay of organic substances in soil, we use the concept of half-life. One half-life is the time required for the decomposition of one-half the mass of the original organic material. This sentence can be expressed mathematically as a process described by an exponential equation. Average half-life values of soil humus manifest very broad ranges, from years to hundreds of years. Their values depend upon the various components of humus. Generally, higher half-life values indicate better qualities of soil humus.

In order to further explain, let us consider the half-life of humus in medium fertile soils developed in a mild climatic zone. Being roughly equal to 100 years, we learn that from each 1 g of humus, only 0.5 g of humus remains after 100 years because 0.5 g of the original humus decomposed. During the next 100 years, the remaining 0.5 g of humus is the source for further decomposition: 0.25 g is decayed, while 0.25 g is left as the source of decomposition during the next 100 years, and so on. The previous two sentences were merely an explanation of the mathematical term half-life. An understanding and appreciation of its real value for humus depends upon the type and quality of the parent organic material, on the environmental conditions for microbiological life, on the soil texture, and generally on the chemical composition of humus.

5.3.3   Humic Substances

We have mentioned the quality of humus, and since humic substances are the main component of humus, we have to explain how to understand their quality. Even if their structure is varied and extremely multifarious, they have several common features. They originated by condensation and polymerization of nuclei formed by cyclical hexagonal arrangements of C–C and C=C bonds or simply by aromatic compounds mutually connected by bonds. Their nuclei carry the reaction groups responsible for the outside charge of humic substance molecules. They are negatively charged since the carboxylic groups –COOH and phenolic groups –OH have the common property of their hydrogen being separated, or as we say, the H is dissociated. We spoke earlier about the H bond and its dissociation. Here it means that the negatively charged oxygen of the mentioned groups (see the missing H) causes parts of humic substances to be negatively charged. Variable arrangements of these basic components result in differences of physical and chemical properties of the humic molecules (Fig. 5.13).

The formation of a humic substance described above can be compared to the steps leading from sheep wool to a hand-tailored suit of clothing. The fine, individual fibers of wool are first spun into spools of yarn that are next woven into the woolen fabric used for sewing the suit. Something could be spoiled on each step starting from the type of sheep and quality of fiber determined by its diameter, crimp, color, and strength and ending with the tailor's skill. A similar change of the quality of the final product of humification could happen. In addition to the condensation procedure, humification brings the products of decomposition into the process, and these products individually persist or more frequently they enter condensation and polymerization, or eventually copolymerization. These steps are responsible for humic substances being a multicomponent mixture of organic polymers. Among them is a small group called glomalins, which originate by decomposition, transformation, and polymerization of decayed parts of plants. They appear as a direct product of fungi living on roots of plants (more about it later on in Sect. 6.2 dealing with soil structure). We should not forget that conditions of soil

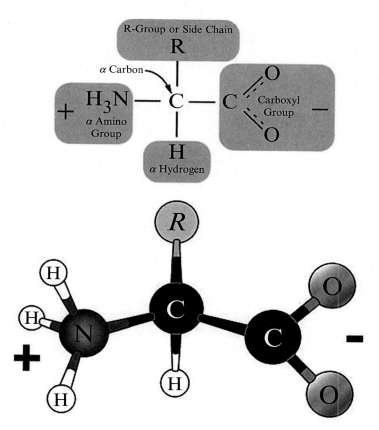

Fig. 5.13 The structure of a humic molecule. The molecule has the positive charge on the side of amino group and the negative charge on the side of carboxyl group. The molecule can be attracted either to negative charges in the soil or to positive charges in the soil in accordance to our simplified description of the humic molecule

environment like pH, aeration, etc. also influence the type of each of the mentioned processes.

Because of the outside portion of humic substances that carry negative charge, they attract and hold positively charged cations like potassium, calcium, and magnesium that belong to important plant nutrients. Their fixation to negatively charged parts of humic substances is not as strong as the bonds inside of the humic compounds. We already described this property in a former section of this chapter as cation exchange capacity. Plant roots are active in this exchange game in order that their roots gain the nutrients needed for their metabolism and growth. But at the same time, nutrients not absorbed by roots are still kept in their same exchange positions on the humic substance and prevented from leaching out of the soil by percolating water. Everything is similar to our shopping for groceries. If we behave reasonably, we usually buy only enough food that we plan to eat at home today and during the next several days in spite of the supermarket overflowing with food.

Another important feature of humic substances is that they form a habitat for microbes and other microorganisms. Some humic materials are much more important than others, while some play a trivial or even a hindering, noncompatible role. Because humic substances form more or less chains and balls on the scale of molecules, they can behave as films that cover or wrap around soil particles or as small balls they can be glued to soil particles owing to various potential bonds. The smaller are the mineral particles, the more intensive is the adhesion of humic substances. When they associate with clay minerals, they exert their maximum capacity to adhere to a mineral surface.

Humic substances are classified into three chemical groups based on solubility and other physical and chemical properties. *Fulvic acids* are the humic fractions that are lowest in molecular size and molecular weight. They have the lowest cation exchange capacity and are lightest in color. Fulvic acids are soluble in both acidic and alkaline solutions and are more susceptible to microbial degradation than other types of humic substances. *Humic acids* are medium in molecular weight and their color is darker. They are soluble in alkaline solutions but insoluble in acidic solutions. They possess an intermediate resistance to microbial degradation. *Humins* are dark colored, have the highest molecular weight and highest cation exchange capacity, are insoluble in both acidic and alkaline solutions, and are the most resistant to further microbial degradation. All three groups are relatively stable in soils. Fulvic acids have the lowest half-life ranging from 10 to 50 years, while the longest half-life reaching up to several centuries belongs to humins. Soils with black color in their top horizons belong to the most fertile soils. Owing to their black color, chernozem is their name (the Russian word for "black earth"). Their humus has high content of humins.

Fig. 5.14 Model of the humic acid as a simplified organic polymer

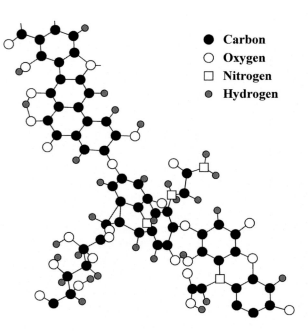

● Carbon
○ Oxygen
□ Nitrogen
● Hydrogen

All decomposition, transformation, condensation, and polymerization processes are also influenced by the mineral part of the soil. Soil particles influence the rate of all these processes. We note that lignins, early stages of humus, are very weakly transformed in sand, moderately in loams, and very intensively in clays. The initial products of early stages of humification are readily linked to clay minerals. Generally we are inclined to say that each of the various structures of silicates is an instructor teaching organic compounds how to form the long-chain linkage by polymerization. Many of these chains are directly joined to clay minerals, while some are enrolled into small balls and connected through the charged parts of the reactive compounds to the crystal lattice of clay minerals. Such linkage assists in slowing down the decomposition rate of humic substances. The clay-humic associations form a very useful habitat for microbial communities since they offer food substrate and protection through a great specific surface in ranges of tens to several hundreds of m^2 per gram of soil. Microorganisms express their gratitude by excreting polysaccharides and proteins that stabilize the organo-mineral associations. The strength of bonding increases with the decreased size of soil particles. The special forms of structure of individual components in humic acids and humins also play an important role – the older and more "mature" they are, the stronger is the bonding. As a result, soil humic substances stabilize organo-mineral associations more in subsoil layers than those associations in the topsoil. All of these aspects are also important in our understanding of soil structure that we will discuss in later chapters (Fig. 5.14).

Chapter 6
Crumbling of Soils

6.1 Soil Structure

Everybody has had many opportunities to observe breadcrumbs or the crumbs of a cake. Within each crumb one can easily see coarse as well as tiny small holes or pores. And between crumbs, the pores are really big. Whenever a bakery produces bread without those big pores, the bread is heavy, has no tendency to crumble, and is hard to swallow. People stop shopping at that bakery. Similarly, cakes without an attractive pattern of large and small holes and crevices are heavy, appear like a piece of concrete, and are unwanted in a bistro. Without satisfied customers the bakery and the bistro are both on a path to bankruptcy.

A good soil should crumble like a good bread or cake. Soil crumbs are called aggregates. They display a very broad distribution of pore sizes ranging from ultra-microscopic pores up to pores recognized by a naked eye. Speaking scientifically, there are pores of cross section from less than μm up to millimeters. Because such a crumbling soil has a special structural arrangement of particles, we speak briefly about *soil structure*. Since there is the similarity with breadcrumbs, we find the term crumb structure in many languages, and this type of structure is considered as the most favorable for plants. The comparison of soil to bread is by no means an essayistic style or manner. Although the term was introduced into the scientific world by the famous Russian soil scientist Dokuchaev in the nineteenth century, he did not invent it. He took it from farmers who distinguished between a good crumb structure of a high-yielding topsoil and a structureless, poor fertility soil. Structureless soils were associated with low harvests and a lack of rye and wheat that often led to famine. But we have to describe now the morphologic features of soils, and therefore we must resist temptation to write an essay on poor structureless soils and famines.

A structureless soil in farmers' jargon does not crumble but forms big clods just after plowing with the clods copying the shapes and numbers of the plow blades. With the surface left idle and no longer tilled, the individual clods typically fracture

© Springer Science+Business Media Dordrecht 2015
M. Kutílek, D.R. Nielsen, *Soil: The Skin of the Planet Earth*,
DOI 10.1007/978-94-017-9789-4_6

and break into smaller amorphous clods as the soil dries. Eventually, the cloddy surface erodes into a rough, shapeless configuration, and after a series of rainfalls and intermittent drying conditions, it is transformed into a nearly impenetrable hard crust. To prepare such a soil for crop production, the farmer follows his initial plowing gesture with additional cultivation procedures specifically chosen to potentially improve the meager tilth of his structureless soil. Even with all of his effort, conditions within the soil are not the best for the growth of roots and generally do not offer an optimal environment for profitable crop production.

Up to now we were dealing with soil structure easily visualized within the tilth or top layer of a soil. A soil scientist's name for soil layers is horizons. The name does not imply that the layers are strictly horizontal – they could be slightly inclined, irregularly tilted, or undulating. But when we dig a pit at one location, we can recognize that soil color, texture, compactness, and the shape of structural aggregates change with depth. By quantifying these morphological features that are typical for various soil types, units of soil structure are identified and named. We shall describe them in the second half of our book.

Now back to our horizons. The crumb structure is typical for the top A horizon (humus horizon). In horizons at greater depths below the A horizon, there are differently shaped structural units of aggregates. In some instances, we find differently shaped aggregates even within the A horizon. Specific types of structure, defined according to the shape of aggregates, may slightly differ between individual national systems. Sometimes the term ped is used for aggregates originating strictly from natural soil-forming processes. At other times, the existence of soil aggregates is attributed to a combination of natural processes and man's activity, example in Fig. 6.1.

If all three main axes of aggregates are roughly equal, the type of soil structure is blocky. In some systems angular blocky structure (faces of aggregates intersect at sharp angles) and subangular blocky structure (corners of faces are mostly rounded) are recognized. If the aggregates are more or less rounded, we speak about granular or crumb structure. Aggregates with horizontal axes shorter than the vertical are typical for prismatic structure. Columnar structure is similar to prismatic structure, but aggregates are rounded at the top. Aggregates with horizontal axes longer than the vertical are characteristic of plates. In some systems this platy structure is separated and called flaky because of the very thin shapes of flake-like aggregates. A honeycomb type of structure is also specified. If aggregates are not formed and individual soil grains are recognizable, then the structure is single grained, like in sand. When no aggregates exist, the texture is loam or clay, and the soil breaks up into big clods, the type of structure is named massive or amorphous. If such a massive or amorphous type of structure exists in the top A horizon, it is identical to a structureless soil. The classification schemes are not rigid and speaking less diplomatically, there is a free space for fabulosity and phantasm.

The degree of structure development and of aggregate stability is its grade. This strength is described simply by words and by numbers from 0 to 3. Grade 0 indicates that no soil aggregates exist, and the soil is structureless. The weak soil structure is denoted by 1, while 2 indicates medium structure. Lastly, 3 is for strong structure. Strong structure can easily be seen in a soil pit, since the soil aggregates

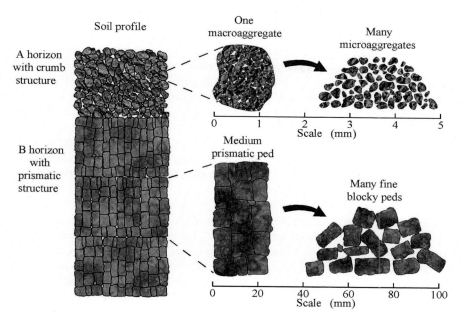

Fig. 6.1 The example of soil structure in the soil type consisting of A, B, and C horizons where A horizon is rich in humus, B horizon originated by slight transport of silt and clay particles from A horizon by percolating rain water. The shape and type of soil aggregates depend on soil evolution and upon man's activity

have large gaps in between the aggregates. Weak soil structure is not easily seen in a soil pit, and aggregates break into a discernible shape when they are manipulated or squeezed. Medium is somewhere in between strong and weak. The determination of the grade is rather subjective, especially grades 2 and 3.

The size of aggregates is not uniform for all types of aggregates. We describe one of possible classification systems in order to show the principle criteria. When all three axes are roughly of the same length and the type is blocky, the size less than 5 mm is classified as very fine, the range 5–10 mm is fine, 10–20 mm is medium, 20–50 mm is coarse, and above 50 mm the type is very coarse. Granular aggregates have substantially smaller boundary values: less than 1, 1–2, 2–5, and 5–10 mm. Very coarse does not exist. The height of platy aggregates is decisive and has boundary values: less than 1, 1–2, 2–5, and 5–10 mm. Aggregates with extensive vertical axes have these decisive heights: less than 10, 10–20, 20–50, 50–100, and more than 100 mm for very coarse aggregates.

The description of soil structure starts by its visual evaluation in the field. A chunk of topsoil extracted with a spade can be manually broken along "fracture" lines into several big clods. With these naturally developed fracture lines being frequently visible, each clod can be repeatedly crumbled to expose a series of smaller and smaller aggregate units. The relative sizes, shapes, strengths, and spatial distributions of all of the aggregate units and clods provide a basis for classifying and naming the observed structure. Such visual field observations are sometimes supplemented with penetration resistance measurements when we determine the pressure necessary to push a cone-tipped rod or pin into the soil. Such measurements are a vast improvement over those used a half-century or longer ago when a farmer pushed his thumb into the soil and subjectively estimated the penetrability of the soil. Today, we have instruments with a scale showing how many Pascals (more precisely MPa or kPa) are applied to push the rod or pin down to a prescribed soil depth. Except for sands, the higher the applied pressure, the poorer is the soil structure. Measurements of penetrability are sometimes accompanied by water infiltration tests that are discussed in Sect. 10.1.1. Many results of laboratory measurements of soil physical properties are also related to soil structure, as, e.g., soil porosity or hydraulic conductivity, Sects. 7.3 and 9.2.

Up to now we described the features of macroaggregates recognized by the naked eye. These large aggregates, composed of microaggregates ranging in size from 10 to 250 μm, are studied on thin sections. An undisturbed small soil block is first saturated with epoxy resin and heated, and after the glue-like resin solidifies, the hard soil block is cut in a special machine. The cut side is polished and placed on a glass slide. Then the soil is cut on the opposite side and trimmed to about 30 or 50 μm (equal to 0.03–0.05 mm). Light from a special microscope can be successfully transmitted through the 30- to 50-micron thin section. Inasmuch as different minerals have different optical properties, most minerals can be easily identified using polarized light. From these thin sections, we can study details of the edges of silicates and their structural arrangement, the forms of Fe and Al oxides and their coverage of particles, or how humus is bound to particles. More detailed studies are possible using an electron microscope to examine the scattering of neutrons or X-rays. Scattering occurs exclusively at the interfaces between the solid matrix and pores containing a different material, such as air or water, which differs substantially in its ability to scatter the radiation. From such microscopic studies, we learn how various arrangements of microaggregates form different sizes and shapes of macroaggregates. And from illustrations of intricate pore space distributions, we readily understand and appreciate the key role of microaggregates as they spatially protect humins and humic acids from microbial decomposition.

6.2 Magic of Soil Aggregation

The first prerequisite for the origin of soil aggregation and soil structure is the existence of an attractive force between soil particles, especially those of clay minerals and soil humic substances. The surfaces of both – clay minerals and humic substances – carry a negative charge. Hence, both attract positively charged ions, the cations from the water solution that always exists in soil pores. If bivalent cations dominate the solution, they are preferably attracted to the negatively charged clay minerals and humic substances causing their concentration near the solid particle surfaces to be higher than that in the solution further from the particle surfaces. The so-called adsorbed cations are so numerous and so close to the particle surfaces that they do not move outside of this domain even if the outside solution flows. The negative charge of the particle surfaces is completely compensated even if the water solution flows through the pores. The particles together with the adsorbed cations form a unit without any charge, i.e., each unit behaves as if it were neutral. When two such neutral units meet, they are not repulsed – on the contrary, they are mutually attracted, form micro-floccules, and coagulate. A typical example of the coagulation-causing cation is calcium, Ca^{2+}.

A completely different behavior is manifested when monovalent cations like Na^+ are adsorbed on the negatively charged surfaces of solid clay particles and humic substances. If they have to balance the same negative charge, their number should be twice that of bivalent cations. Let us give an oversimplified example by considering that a very small solid surface has 100 negative charges. To balance these negative charges to obtain a charge of zero, the condition for flocculation, we have to supply the adsorption zone with 50 calcium cations (Ca^{2+}) or with 100 sodium cations (Na^+). In reality, because there is usually not enough space in proximity to the solid surface to accommodate 100 monovalent cations, a certain percentage of those cations remains outside of the unit (particle + adsorbed cations) when soil water flows. As a result, the unit (particle + adsorbed cations) retains a negative charge. Again with oversimplification: if there is only space for 80-monovalent-cation juxtaposition to the surface of the solid particle while water is flowing, 20 cations move away with the water leaving 20 unbalanced negative charges. Therefore the unit (particle + adsorbed cations) maintains a portion of the negative charge of the solid particle. Although it is less than the 100 negative charges of the solid surface without adsorbed ions, the unit nevertheless sustains its negative charge behavior. Hence, when two such negatively charged units meet, they repulse each other. Without being mutually attracted, they cannot flocculate to form the originating nucleus of a future microaggregate. In addition to the above-simplified counting of charges, there is another factor why the majority of monovalent cations cannot be pushed into the film around the particle and coexist with it. That factor is the motion of water molecules rotating and orienting themselves in the vicinity of an ion in such a way to form an envelope of water continually surrounding the ion. Because the thickness of water envelopes around monovalent cations is generally much greater than those around bivalent cations, the size of a sodium cation together with its larger water envelope is much bigger than a calcium cation with its smaller water envelope. Hence, there are two primary factors that prevent soil particles with adsorbed monovalent cations from coagulating and forming a nucleus of microaggregates. The most active in this negative activity is the sodium cation, Na^+. If the water solution is changed and Na^+ starts to prevail, the earlier stability of microaggregates is lost since an opposite process to coagulation starts, and the clay particles are repelled.

As we have already mentioned, cations attracted to soil particles from solutions of mineral salts in soil are readily exchanged when the composition of the soil solution changes. With this change of solution composition, basic soil properties also change. One of the sources of dynamics in microaggregation is the direct consequence of this inevitable change in soil properties.

As plants grow, their fine roots penetrate through the soil by pushing and tearing it into many small fragments separated by a network of pores filled by roots and root hairs. When the plants expire and their network of roots starts to decay, they bequeath an inheritance of a crumbled soil. A certain portion of the root organic matter is transformed into humus that cements coagulated floccules and small fragments together. The humins and humic acids serving as the glue are resistant against the action of water, especially against their dissolution in water. For a long time in the past, it was assumed that the content of all humified soil organic matter had this positive effect: the more of it in soil, the more advanced is the existence of aggregation. During those times it was taken for granted that humins played the decisive role regarding the quality of soil aggregates. When we speak about quality of aggregation, we must first accept the necessity of virtually all components of soil being transformed into aggregates of long-lasting stability and then base our judgment on their capacity to retain plant-available nutrients, including water.

A special group of molecules detected during 1996 has an action on the quality of aggregates that is several times stronger than that of humins. These molecules got the name glomalin since they have been produced by symbiotic fungi *Glomales* (order *Glomerales* according to the International Botanical Taxonomy). Glomalin consists of special types of proteins, and because of their unique structures, they are called glomalin-related soil proteins. Since this term is too long and complicated for many nonspecialized readers, soil researchers use the abbreviation GRSP. Proteins are polymers built from alpha amino acids in the form of a chain. All alpha amino acids consist of alpha carbon, i.e., the carbon that is attached at the first (alpha) position, to which an amino group (H_3^+N) is bonded together with carboxyl group (COOH) and side chain which is very variable in chemical structure. The chain is three-dimensional like a wide ribbon twisted into a spiral with a variety of bonds. These variations result in plurality of properties.

Fungi *Glomales* have occurred in mycorrhizal symbiosis with plants when the green plants entered into terrestrial conditions in Silurian and Devonian about 400 million years BP. The name of the symbiosis is derived from the Greek *mykos*, meaning fungus or mold, and *rhiza*, meaning root. These fungi exist in about 80 % of all terrestrial vascular plants and in some mosses. They are branching like a tree on the microscopic scale, and since the Latin *arbuscula* denotes also very small *arbos* (tree), we define the symbiosis between those fungi and plants as arbuscular symbiosis. It is known that fungi penetrate into fine hair roots of plants without damaging their cells. The long fine outside fibers of *Glomales* fungi bring water and plant nutrients into roots, especially phosphorus and trace elements from the soil. Because the external fungi fibers occupy a much bigger space of soil than the roots, they pump much more water and elements into plant roots than the root system is able to achieve. The plant acknowledges this beneficial gesture by supplying photosynthetically produced organic compounds to that portion of fungi fibers inside its roots. With the fungi receiving mainly glucose and similar metabolic compounds, the plants and the *Glomales* both profit from this symbiosis. The surface of the external fungi fibers is reinforced and fortified by molecules of glomalin that are produced in the body of fungi from organic compounds delivered by the plant. After a couple of weeks when the fine roots have matured, the external fungi fibers cease to function, the molecules of glomalin enter into the soil, and the second period of glomalin action begins. Glomalin's primary aim was the reinforcement of the external fungi fiber walls, and when this aim disappears, glomalin remains in the soil applying its fixing property upon its next partner – the soil particles. With the half-life of glomalin being at least 40 years, many reports of research frequently state that it continues to exist for about 100 or even more years. Its resistance against decomposition is much higher than that of humins.

Various reaction groups on glomalin's molecular surface account for its tenacious ability to glue fine silt particles and clay minerals together. Moreover, it makes linkages with other organic molecules. Through these and other attributes, glomalin becomes a very active glue to bind microaggregates together to form macroaggregates. Historically until the very end of the twentieth century, the presence of glomalin and its advantageous binding contributions to aggregation remained a mystery.

We now know that glomalin was discovered 17 years ago by Sara F. Wright, a scientist of the USDA Agricultural Research Service. Soon after its discovery, many research papers were published on aggregate fixation by glomalin when changes occurred in land use or tillage technology. With the dominant role of glomalin in soil aggregation and soil protection being confirmed, it is now beyond any doubt that glomalin is much more important than humins and humic acids for maintaining and sustaining aggregate stability.

Here, we must remind ourselves that earthworms also play a vital role contributing to the existence of soil aggregates. Taking a rough average of data, one earthworm swallows in one day the same weight of soil as is its own weight. Their excrements are first-class aggregates. We should also remember to honor Gilbert White (1720–1793) who documented the unique contributions of earthworms to soils more than a century before Charles Darwin's historical books. At that time, concepts of soil fertility were at their infancy and soil structure was not known.

Historically, man's management of soils for agriculture has generally had a bad influence upon soil structure. To his credit, he sometimes practiced a fallow system when the land was neither plowed nor used for one or more years to enhance the fertility of his farm for the next time that he tried to grow a crop. We know now that the soil fertility was enhanced mainly by the restoration of humus content including the rise of glomalin. In such cases, the deteriorating crumbled soil structure was strengthened and the accessibility of plant nutrients was renewed. On the other hand, owing to financial pressures and economics, fallow periods were often ostracized with the frequency of cultivation intensified. This intensification gradually decomposed both humus and glomalin without sufficient opportunity for their replacement. The quality of soil structure decreased and machinery used for tillage, harvesting, and transport of harvests finished the act of soil structure destruction. The compacted soil with its pseudo-aggregates and clods was further damaged by erosive rains when water did not infiltrate into a muddy, sludgy surface soil. When the soil dried out, a compacted crust formed on its surface. All these negative consequences have reduced yields and deteriorated the ecologic environment (Fig. 6.2).

MACROAGGREGATE

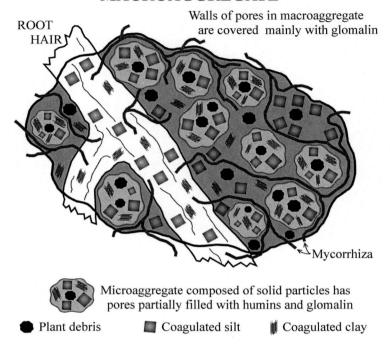

Fig. 6.2 The role of humins and especially of glomalin upon aggregation

Recent research has revealed that the content of glomalin decreases more gradually when no-tillage technology replaces classical tillage and when crop residues are left in place as a source for humification. However, each advantage is usually linked to a disadvantage – in this case, a proliferation of weeds. Although applications of herbicides prevent weed growth, they endanger the quality of groundwater as well as water in rivers. Farmers are now finding themselves in the middle of another paradox when they use fungicides to control troublesome crop plant diseases (molds, mildews, and smuts). With fungicides killing *Glomales*, the fungus that produces glomalin, the final result would be that farmers would strive to make a living growing poor-yielding crops on a structureless soil. Here a lot of research needs to be done.

Chapter 7
Soils as the Skyscrapers

Whenever we observe a skyscraper, we immediately recognize an extremely tall building usually made of concrete, steel, and glass towering above nearby buildings. Our experience assures us that inside the huge structure, ample space provides a healthy environment with human activities pulsating normally owing to sophisticated forms of machinery operating within the remainder of the skyscraper's internal structure. On the other hand, when we look at a soil clod, we are completely unaware that we have an ultramicroscopic skyscraper in our palm. Thinking of the clod as a solid mass, we cannot imagine it as an ideal home for organisms living within its cavities. It never occurs to us that its cavities, occupying about one-half the volume of the clod, provide a healthy environment for its huge number of tiny animals and plants living together.

7.1 Soil Pores

The space surrounded by individual solid soil particles is called a pore. Provided it does not contain any solid soil material, such a cavity is designated as a pore regardless of its size and shape. Pores exist as coarse channels after roots have decayed and as fissures after soils containing a high percentage of smectites become dry. Moreover, pores exist within and between soil aggregates.

The relative volume of all pores in a specific volume of soil is defined as soil porosity with symbol P and frequently denoted as percentage. If we write $P = 45$ %, we mean that a 100-cm^3 volume of soil contains 45 cm^3 of pores and the remaining volume of 55 cm^3 is filled by solid soil particles of either inorganic nature or organic origin. Porosity is sometimes indicated by a decimal rather than a percentage. In our example, its value would then be 0.45.

Soil pores are frequently denoted as voids in studies of mechanical properties of soils. If the volume of voids is divided by the volume of solid particles in the soil, we obtain the void ratio e, expressed as a decimal. For a clear illustration, we write

© Springer Science+Business Media Dordrecht 2015
M. Kutílek, D.R. Nielsen, *Soil: The Skin of the Planet Earth*,
DOI 10.1007/978-94-017-9789-4_7

here a few mutual relationships: for $P=0.4$, $e=0.67$; for $P=0.45$, $e=0.82$; for $P=0.5$, $e=1$; for $P=0.55$, $e=1.22$; etc. In well-graded dense sand, $e=0.45$ (related to $P=0.31$); in loess frequently in ranges $e=0.85$ to $e=0.92$ ($P=0.46$ to $P=0.48$); and in the top A horizon of the majority soils, we find e in ranges between 1.0 and 1.25 (P between 0.5 and 0.55).

Soil pores can be demonstrated or modeled in various levels of approximation. An oversimplified arrangement consists of spherical particles having the same size, i.e., the same radius. With each particle contacting a neighboring particle at only one point, the particles can be arranged in one of three models (Fig. 7.1).

The loosest arrangement is the cubic model. When we connect the centers of neighboring spheres in such a model, we obtain a cube with each particle having six contact points with its neighbors. When we connect the centers in a two-dimensional

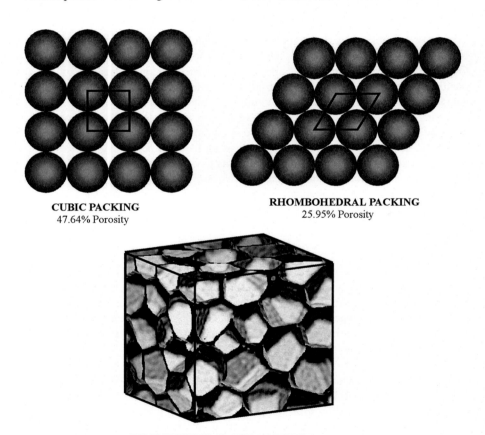

CUBIC PACKING
47.64% Porosity

RHOMBOHEDRAL PACKING
25.95% Porosity

RHOMBOHEDRAL ARRANGEMENT
Equal-Sized Soil Particles

Fig. 7.1 Spherical soil particle models of pores: cubic packing with porosity $P=48$ % (*top left*) and rhombohedral packing with $P=26$ % (*top right*). Rhombohedral packing of more realistic soil particles of roughly uniform size (*bottom*)

arrangement, we obtain a square and four contact points. The porosity of this loosest type of arrangement is 47.64 %, rounded to 48 %.

The tightest arrangement is hexagonal since the connecting of the centers of spheres surrounding a central sphere forms a hexagon in a two-dimensional array provided that the connections passing the central sphere are not allowed. Or, if the centers of spheres are connected in one direction only, we obtain a rhombus, and the model is sometimes called rhombic or rhombohedral. The angle of each rhombus is 45°, not 90° as in cubic packing. The porosity of this tightest packing is 25.95 %, rounded to 26 %. This extremely low value has never been reached in soils.

The third arrangement is a combination of the two abovementioned schemes that generate porosity values between 26 and 48 %.

However, we know that no soil is a conglomerate of particles having identically the same size. In reality, smaller particles penetrate into voids between bigger particles (Fig. 7.2). And instead of being arranged in a regular system of cubes or rhombi, they form all sorts of different kinds of lumps and bridges filling gaps between larger particles. Previously, we implied that isolated soil particles rarely exist. Usually, particles have tendencies to lump together to form microaggregates and, subsequently, coalesce into still larger clusters known as macroaggregates.

Although some models of spherical particles can yield commonly observed soil porosity values of 40–60 %, their design and numerical calculation are extremely complicated. Generalizing such models of pores to the space between spherical particles does not help too much because they fail to reflect the various shapes of soil particles that dominate many important properties of soil pores. We initially selected spherical particles merely to create a relatively simple demonstration of porosity.

In order to overcome the difficulties in applying spherical models to real soils, the pores have been modeled as tubes of various radii. Such models have been found

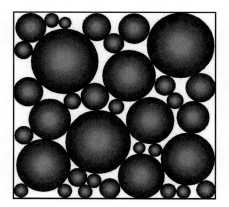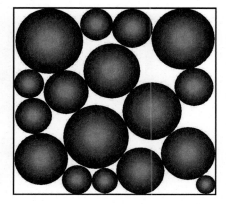

Fig. 7.2 Two illustrations of packing different distributions of unequal-sized spherical soil particles in models that yield more realistic porosities of natural soils

to be successful and easily applicable for derivation of theories on flow of water and solutes in soil. We shall use one of them in the next two chapters (Fig. 7.3).

We should remember that the above models – one composed of spheres and the other being an assembly of very fine tubes – are still very distant from reality. Nevertheless, they continue to be presented in textbooks for explaining basic ideas about soil pores and their role in the majority of processes that occur within soils. A somewhat better model exists when tubes having variable radii along their lengths are connected by spheres (Fig. 7.4).

Fig. 7.3 Pores modeled as an ensemble of parallel cylindrical tubes with each having a different radius

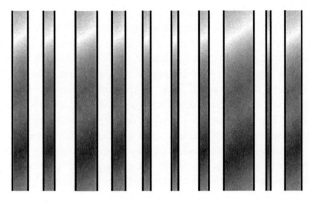

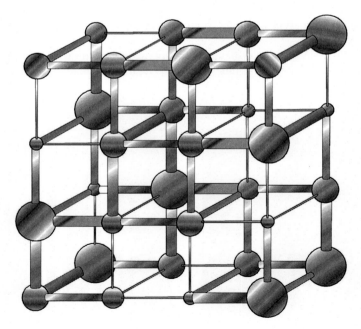

Fig. 7.4 Model of soil porous system consisting of a combination of spheres and tubes, both of which vary in size

Let us abandon for the time being the complicated description of pores and their natural arrangement within soils. Dealing with soil porosity is much more practical because its determination is relatively simple. Examining changes of soil porosity owing to existing vegetation and man's activity has been the aim of many research studies during the last two centuries. This accumulated knowledge of porosity is therefore much more important in our practical life than studying the geometry of pores. Hence, we shall spend more time and space on porosity than on approximate studies about soil porous systems. Our intention is not to underestimate the utility of soil porous system theories – to the contrary, where we find the opportunity to show illustrative results, we use them without going into detail of unnecessary advanced theories and methods.

7.2 Soil Porosity

The volume of pores V_P related to the total volume of soil V_T in its natural configuration of soil particles is denoted as porosity, as we have written in the previous sub-chapter, $P = V_P/V_T$. Before we measure its parameters, it looks like a very simple task until we have to decide exactly how large a volume of soil V_T should be measured. How we make this decision is illustrated in Fig. 7.5. Notice in the top part of the figure that when we initially choose a very small volume V_{T1} (e.g., the size of a sand particle), everything depends upon the position of the center of the volume we sample. If a sample is taken from the center of a particle, we obtain porosity $P = 0$. If a sample is taken from the center of a big pore, we obtain $P = 1$ (i.e., $P = 100\%$). If we decide to sample a bigger volume V_{T2} with its center located on a solid particle, we determine $P = 0.33$. Sampling the bigger volume V_{T2} with its center located on a pore, we determine $P = 0.72$. For a still larger volume V_{T3}, $P = 0.40$ (centered to solid particle) and $P = 0.49$ (centered to the pore). If we proceed to systematically increase the sampling volume for two cases (1. centered on a solid particle and 2. centered on a pore), we obtain two curves that differ substantially for relatively small sampling volumes; see the graph at the bottom of Fig. 7.5. As the volume of sample continues to increase, the two curves finally merge to the same value of porosity P regardless of where the sampling was centered. This volume is the representative elementary volume, REV, and is theoretically applicable to exact studies within prescribed limitations. In practical applications, decades of experience have proven that a volume of about 100 cm^3 is sufficiently adequate and precise in the great majority of instances.

As a result, the determination of soil porosity is rather simple. We sample a natural field soil using a metallic, cylindrical device having a diameter and height each equal to about 6 cm. For special circumstances, the height is adjusted in accordance to the aim of the study.

We push the device into the field soil, remove the soil sample of known volume, and oven-dry it in the laboratory. Weighing it after it becomes dry, we calculate the soil bulk density. We next measure the specific density of the soil particles – another

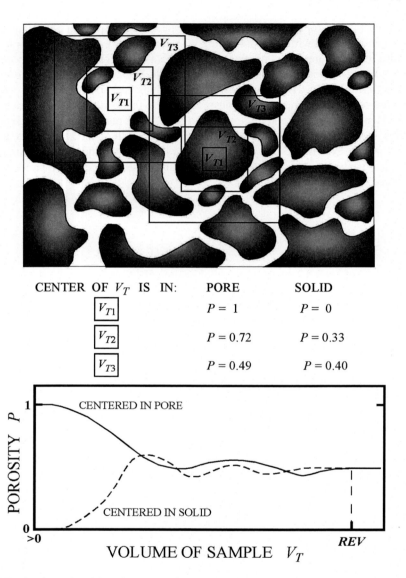

CENTER OF V_T IS IN:	PORE	SOLID
V_{T1}	$P = 1$	$P = 0$
V_{T2}	$P = 0.72$	$P = 0.33$
V_{T3}	$P = 0.49$	$P = 0.40$

Fig. 7.5 The size and position of an experimentally measured soil sample should provide identical estimates of porosity P for the representative elementary volume *REV* of the soil

simple task. A special glass container with a glass plug, called a pycnometer, is filled with water and weighed. Next, after a known weight of the dried soil is placed into the pycnometer, it is filled to its capacity with water and weighed again.

From the first and second measured weights, we obtain the particle density, which is about 2.6 g/cm³ in the majority of topsoils containing significant quantities of humus. Soils from deeper horizons without humus have slightly higher values, usually close to 2.7 g/cm^3. The porosity is simply calculated from the two densities.

Because individual pores in sandy soils can easily be observed with a naked eye, it may seem as a surprise that sandy soils have the lowest porosity of all soils. The arrangement of their individual particles resembles a combination of cubic and hexagonal packing held together by small admixtures of silt and clay particles. Their porosity, about 40 %, may increase with increases of humus content, but it never reaches the values of loams, clays, or other textures. The porosity of loamy soils usually has a larger range of 52–64 %, because pores exist within three different sized domains – between individual particles, within microaggregates, and within macroaggregates. With the development of more aggregates, the improved soil structure increases the total porosity. Even when the soil structure of a loamy soil looks somewhat deteriorated or even destroyed, its microaggregation usually persists to maintain its total porosity above that of sandy soils. The porosity of a loam falls below 50 % whenever it is slushed after a rain or is more or less permanently waterlogged. If this structureless puddled loam dries out on the soil surface, it keeps its low porosity and forms a solid skin of topsoil. Clays and loamy clays usually manifest a lower porosity than that of loams. However, the drying of structureless clays and loamy clays is accompanied by shrinkage and the formation of fine and coarse cracks causing an increase of porosity sometimes above the values of a loamy compacted skin on the surface.

When we dig out a clod of fresh, moist soil and squeeze it in our palm, we find in addition to its changed shape that the overall size of the squeezed lump is smaller. This reduction occurs because we compressed and transformed the biggest pores into smaller pores. Consequently, the size distribution of pores can be decreased when a soil is compressed, e.g., by traffic or by heavy agricultural machinery and vice versa.

7.3   Soil Pores Like a City Center

The variable shapes and size of soil pores continually change. It is not uncommon for a pore of 1 mm in cross section to be drastically reduced to merely a micrometer (1 μm = 0.001 mm). The biggest pores display more or less parallel walls with distances up to 2 cm and are called cracks. Earthworm channels have the shape of irregularly curved tubes with diameters of 4 mm or less. Pores originating after the decay of grass roots have diameters ranging from 0.1 to 0.3 mm. Both are sometimes called biopores together with all pores originating under a direct and immediate influence of microfauna and flora. Pores between aggregates have most frequency diameters between 10 and 200 μm, and in loose well-aggregated soil, they have diameters up to 1 mm. Pores inside aggregates are substantially less – only 0.2–10 μm. All of the pores between aggregates are sometimes denoted as macropores and those inside aggregates are micropores. The range of pore size is very broad in soils; see Fig. 7.6.

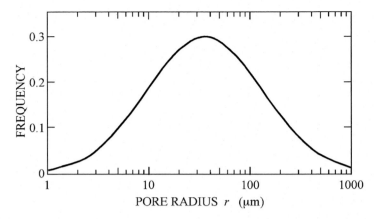

Fig. 7.6 A simplified, ideal shape of a lognormal pore-size distribution in a structureless soil. The horizontal axis is in the logarithmic scale. The peak of the curve indicates the most frequently occurring pores (or cylindrical tubes of models). They have radii about 30–50 μm

Because the surface of pores is not smooth, thin sections of soil observed by microscope offer us fantastic images of pore formations. Sophisticated procedures allow us to transform the images into quantitative analogues of geometric arrangement and to better understand the term *diameter* used in the previous paragraph.

Let us first simplify the shape of soil pores into vertical cylindrical tubes of radii r ranging from 1 μm (micrometer=0.001 mm) to 1 mm. Next, if we arrange them increasing from their smallest size ($r = 1$ μm) on the left to their largest ($r = 1$ mm) on the right, we obtain a picture similar to the pipes of an organ. Designating the length of each tube of radius r according to its percentage occurrence in the soil, we obtain the relative number of tubes, i.e., their frequency. Plotting their frequency on the vertical axis and their radial size in logarithmic scale on the horizontal axis, we obtain a curve similar to a bell as illustrated in Fig. 7.6. Most frequent are the pores of radius 30 μm in our example, and the peak of the curve depends mainly upon the soil texture. Lowest frequency belongs to the thinnest tube ($r = 1$ μm) residing at the left-hand side of the graph, while the fattest tube ($r = 1$ mm) having nearly the lowest frequency is at the far right-hand side of the graph.

Since the soil pores have irregular shapes, we compare the behavior of soil water and flow of liquid through the soil to that through the parallel tubes in the model. This parallel tube model provides a convenient means to easily explain the overwhelming complexity of a soil by interpreting the radius of a tube as an equivalent pore radius or simply pore radius.

We owe the readers an explanation of why we plot the pore radius in logarithmic scale – the decadic with log 1=0, log 10=1, log 100=2, etc. Because $10^0 = 1$, $10^1 = 10$, etc., it follows that the length between 1 and 10 μm on the horizontal axis is the same as that between 10 and 100 μm and also between 100 and 1,000 μm. We have the opportunity to follow the changes of the shape of the curves if the peak of two studied curves is in the range of one-half of the order of magnitude as, for example, in Fig. 7.6. Even more important is the fact that the pore-size distribution in real soils is close to the ideal shape in Fig. 7.6, and this "bell" shape is well described using mathematical statistics. If we had used a linear scale where the distance between 0 and 10 is the same as between 100 and 110, then the peak of the pore-size distribution curve would be shifted to the left making differences between the two curves that we present next in Fig. 7.7 less distinct.

A graph of pore-size distribution within a structured soil is more complicated. Its curve plotted with logarithmic scale on the horizontal axis looks like a Bactrian camel with its characteristic double humps. The pore-size distribution curve of a structured soil also has two humps – one representing pores mainly between the aggregates and the second of finer pores residing inside aggregates and between

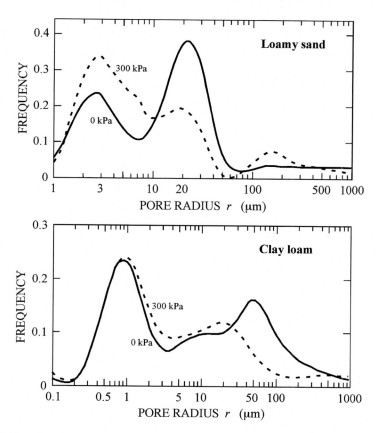

Fig. 7.7 Pore-size distributions in two aggregated soils before and after compaction by an applied pressure of 300 kPa. Before compaction, the most frequent pores between the small aggregates of the loamy sand have a radius of about 20 μm, while those between the aggregates in the clay loam are more than twice as large with a radius of 55 μm. The radius of the most prevalent pores between the particles inside the aggregates of the loamy sand is 2.5 μm, while that of the clay loam is three times smaller, i.e., $r = 0.8$ μm. The broken curves for each soil manifest significant changes of relative pore sizes after a long-time compression with 300 kPa

individual soil particles. Such pore structures for a topsoil of loamy sand and of clay loam are revealed by the solid curves in Fig. 7.7. A peak of the most frequent pores between the aggregates of the loamy sand occurs at a radius $r = 20$ μm, while the most frequent pores inside its aggregates have a peak at $r = 2.5$ μm. Within the clay loam the equivalent peaks occur at radii of 55 and 0.8 μm, respectively. We note that these broad ranges of double peaks are caused by the type and quality of soil structure coupled with variations of soil texture. In addition, soil fauna and decaying roots and other deteriorating parts of flora influence the shape and maximal frequency of soil pores in each of the two main components of the soil pore system. The differences of the two soils are completely obvious when the log scale is used for the equivalent pore radius. If we had used a linear scale, the first peak of micropores 2.5 μm (loamy sand) and 0.8 μm (clay loam) would merge, and the difference between pores inside the aggregates would have disappeared.

Pore-size distributions within soils are always sensitive and vulnerable to the impacts of soil tillage and compression by machinery. Returning to Fig. 7.7, we note that the solid curve for each soil is also associated with a dotted curve derived from measurements of porosity after each soil was individually subjected to a pressure of 300 kPa for a long time. The dotted curve for each soil is drastically changed from its original uncompressed solid curve. The shifting of peaks and the changing magnitudes of pore sizes are not the same for these two soils or for any other soil. In other words, quantitative predictions of changing pore-size distributions remain a challenge for future research.

Under the optics of a microscope, we see large pores existing next to narrow pores, long pores ending abruptly, long pores merging to form big pores, big pores connecting with tiny tubes, and all combinations of different sized tubes leading to countless directions – collectively, we are inclined to see disorder or complete chaos. It reminds us of a map of the center of a big city. There, the main traffic, administration centers of powerful companies, and top-quality shops are concentrated on the main avenue. Next to it are ordinary streets and narrow alleys and passageways leading one to another, while some passageways are blind or narrowed just for foot passengers. We could say again, what a chaos! But it developed with several principal aims: to enable the most efficient transport of goods as well as of citizens. This network offers a place for business, administration, destructive elements and law enforcement, and, even more generally, all important functions of the city. In a similar way, the soil pore system looks chaotic, but it evolved to offer an optimal network for the life and development of the type of soil we are observing and soil scientists are studying.

When we compare all models to the reality that we attempt to study on thin sections cut from natural soil, we readily acknowledge that the models are crude estimates regarding the variance and distribution of spatial pores. But even so, the models help us understand the measured physical and chemical characteristics and processes taking place within soil pores (Figs. 7.8 and 7.9).

Fig. 7.8 Various shapes and sizes of soil pores observed in *white* from a microscopic thin section

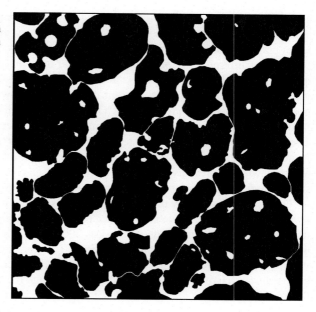

Fig. 7.9 The real configuration of soil pores in three dimensions (*gray*) while the cut pores by a horizontal plane are in *black*

7.4   Carnival in Soil Pores

Soil pores are usually filled partly by water and partly by air. Waterlogged soils having each and every pore filled entirely with only water are truly exceptional cases. The composition of soil air differs from that of the air in the atmosphere. First of all, with soil air having a very high water vapor content, its humidity is always close or equal to 100 %. An often-occurring practical consequence is the high air humidity and undesirable moisture conditions in cellars whenever the basement of a building is not perfectly isolated from the surrounding soil.

Another difference in composition of soil air is its increased content of CO_2 frequently accompanied by a decrease in oxygen concentration. Although this relation was earlier not well traceable, present-day scrutiny shows increased interest. The possibility that extreme increases of CO_2 concentration in soil air could lead to less favorable conditions for agricultural plants has not yet been satisfactorily resolved. The other aspect frequently repeated by climatologists is the role of CO_2 emissions from the soil upon recent climate change. However, we do not consider this hypothesis to have been properly developed, nor has it been checked, proven, and demonstrated by full-size experiments performed by nature itself.

Theoretically we know that climate change is caused by the variation of at least seven factors. If we exclude those acting on a very long time scale, we meet four of them acting on the time scale of the last 200 years. They are (1) changing solar activity, (2) fluctuating geomagnetic activity, (3) rearranging sea streams, and (4) varying concentrations of greenhouse gases, mainly of CO_2. We know more or less exactly how each of them could cause climate change if each of the remaining three factors is kept without change. When we plot such a graph, we obtain, e.g., a rising curve between climate change and each isolated factor as a function of time. When each is plotted we obtain $I(1)$, $I(2)$, $I(3)$, and $I(4)$. However, if two or more factors are changing at the same time, we are not allowed to simply add them together, such as $I(1)+I(2)$, since process (1) influences process (2) and process (2) influences process (1) according to the rules governing a domain in irreversible thermodynamics. In most complicated situations, it could happen in climate change that process (1) will influence all three remaining processes. We are inspired by the Onsager's reciprocal relations in irreversible processes. This theory says that when two or more irreversible transport processes take place at the same time, the processes may interfere with each other. In analogy, the same is valid according to Onsager for chemical processes. Above, we were extending the theory to processes in nature like climate change, and just addition of curves in graphs is not valid. Simply saying, processes in nature are a lot more complicated, and in majority of instances all our recent models are limping on both legs, and we are not allowed to derive the change of climate just from the change of CO_2 concentration. This is our theoretical objection and this is also our explanation why up to now the predictions differ from the measurements.

Our own experimentation with the planet Earth and its atmosphere is not possible. But we do appreciate nature itself for providing evidence of climate change and the concentration of atmospheric CO_2. Just for briefness we restrict our discussion to Holocene (the last 14,000 years) temperature and CO_2 concentration data obtained from ice samples gained by drilling in Antarctic glacier stations. There were about nine cool and nine warm periods. Similar, usually less frequent climate fluctuations were ascertained for other regions, and in many instances, the change of CO_2 concentration was not the cause of climate change but was merely the consequence of climate change that had arrived earlier.

Karl Popper, regarded as one of the greatest philosophers of science during the twentieth century, said that a theory or hypothesis cannot be proven true if found as false in only one instance or situation. Such a hypothesis has to be modified or even rejected. A single falsification annihilates an existing hypothesis even if held or supported previously by the majority of scientists. Hence, we are confident that a lot still has to be done to quantify to what extent the atmospheric CO_2 concentration contributes to climate change. Moreover, until confronted with more convincing scientifically accepted evidence, we steadfastly believe that atmospheric CO_2 has not played a dominant role in global warming during the previous 200 years.

Let us now continue without saying that studies about CO_2 concentration in soil pores and its transport from soil to the atmosphere are useless. Quite opposite, the CO_2 concentration and flux through soil to the outer atmosphere is a good indication on humification.

The majority of biological reactions in soil rely on the presence of oxygen, e.g., the ever-present decomposition of organic matter and the production of soil humus use oxygen and produce carbon dioxide; see Fig. 5.12. Although CO_2 penetrates and moves up by diffusion through the soil pores into the atmosphere, the conditions for diffusion in soil are less favorable than those in the atmosphere. Consequently, because the sequestration rate of CO_2 from the soil is smaller than its production rate, the CO_2 concentration within soil root zone is higher than in the atmosphere. This substantially increased CO_2 concentration increases the concentration gradient – the driving force for diffusion. After reaching this increased diffusion rate, a new balance is established between the production and release of CO_2 from the soil into the atmosphere. If the rate of CO_2 production subsequently increases again, diffusion also increases owing to the increase of CO_2 concentration in soil air and vice versa. The rate of organic matter decomposition depends on many factors – the amount of decayed parts of plants, temperature, humidity and soil water content, degree of pore connectivity, and atmospheric pressure. Inasmuch as these factors are not constant in time, we expect continual fluctuations of CO_2 concentration in soil pores and observe persistent changes of the rates of CO_2 release from the soil. A very rough indication on the CO_2 concentration in the air of soil pores is about ten or more times higher than that in the outside atmosphere above the soil surface. Furthermore, because the value of CO_2 concentration fluctuates drastically throughout the soil profile, the CO_2 concentration at some depths of subsoil could reach even 30 times higher than its value in the atmosphere. The global CO_2 concentration in the atmosphere is now close to 390 ppm, while its value in the pores of the soil humus horizon fluctuates in a broad range of about 4,000 ppm. The symbol ppm means the Latin *pars per million*, part of one million in volumetric units. The value 390 ppm CO_2 is 0.039 % volumetrically or 39 ml of CO_2 in 1 l of air.

The ten times greater concentration of carbon dioxide in the soil compared with that in the air above the soil surface is the essential ingredient to overcome obstacles preventing diffusion, namely, those caused by dead-end and chaotic soil pores as well as the relatively small volume of air through which gaseous transport must occur within field soils. Just for our imagination, let us consider two 100-l barrels filled with water. We leave the water level in the first barrel free to the atmospheric pressure of 1 bar (or 1 atm). After enclosing the top of the second barrel, we use an air pump to maintain a pressure of 10 bars on its water-filled surface. Next, we open the taps on the bottom of each of the barrels at the same time. Everybody knows from experience that the water flows faster from the second barrel that will be empty much earlier than the first barrel. A similar mechanism acts in diffusion.

Now we consider a more exact explanation. Let us imagine that 10,000 molecules of CO_2 are in the topsoil immediately within the surface of a field soil. Above the soil surface there are only 100 molecules of CO_2 in the air. Owing to the thermal oscillation of the CO_2 molecules, there is a high probability that 80 molecules shall penetrate from the soil pores into the atmosphere above soil surface, while at the same time, there is an equal probability that only two molecules shall penetrate from the outside atmosphere into the soil pores. These simultaneous penetrations exemplify a very simplified diffusion of CO_2 in both directions – out from the soil to the atmosphere and from the atmosphere into the soil. The net result is the escape of 78 molecules of CO_2 from soil to atmosphere. In the nineteenth century and first half of the twentieth century, this diffusion process involving CO_2 was frequently called soil respiration. It was not an appropriate term inasmuch as respiration is a mechanically active process, while diffusion through the soil is a passive process. In our simple example had the CO_2 concentration in soil pores been higher, the net flux of CO_2 from the soil would have also been higher. From a practical consideration, we know that whenever the decomposition of organic matter is more intensive than the release of CO_2 into the atmosphere, the concentration of CO_2 increases within the soil profile. On the other hand, when the CO_2 production rate equals the CO_2 diffusion rate, we observe a so-called quasi equilibrium. The term quasi is used since the rates of production and rates of diffusion change instantaneously – each at values differing from the other.

The decomposition rate of organic matter depends upon the amount of water within a soil, e.g., after a rain, an initially dry soil becomes wet and subsequently releases more CO_2. The rate of decomposition also increases whenever the soil temperature gets higher. And because the production of CO_2 decreases when the temperature falls down, the sequestration of CO_2 from the soil is smaller at night than during the day.

Diffusion is also influenced by the content of water in soil. The higher the soil water content ("soil moisture"), the smaller is the space through which CO_2 molecules must travel during diffusion. The conditions for diffusion are relatively favorable if air fills at least 30 % of the soil pores. When the fraction of soil pores filled by air is only 10 % or less with water filling 90 % or even more of the pores, the free connection of soil air stops with air existing in soil only as isolated bubbles. The concentration of CO_2 increases inside those bubbles at the expense of oxygen, i.e., the concentration of O_2 decreases. If the soil is fully or nearly fully saturated by water, i.e., if the soil is waterlogged for a long time, the absence of oxygen leads to a reduction of oxygenation and to the dominance of chemical reduction processes. After extremely long times of soil waterlogging, oxygenation stops completely. Because such conditions are unfavorable for a great majority of cultivated plants, we shall later on describe their consequences in more detail. Now, just for a brief mention, we must say that the decomposition of organic matter is slow and produces alcohols, aldehydes (e.g., methanal, ethanal (or acetaldehyde), propanal, 2-methylbutanal), and other compounds that are generally toxic to the roots of most cultivated plants. Additionally, free iron and manganese as well as their soluble forms released from inorganic compounds and minerals are detrimental to the physical and chemical properties of soil and to plant roots. Moreover, the soil reaction is changed with its pH indicative of acidification. Cultivated plants suffer from these unfavorable conditions stemming from decreased diffusion rates. If diffusion remains nil, conditions eventually deteriorate and lead to the destruction of all cultural plants.

Although the term or expression soil pH is commonly used to designate a specific property of an entire mass or volume of soil, its use actually only characterizes the reaction expressed as pH of the water solution contained in soil pores. Even this restriction is not an exact description of what we measure since we mix the soil with an excess of pure water. The reader already experienced in the chemical term pH can skip the following two paragraphs because we next explain pH in a popular manner.

A very small portion of water molecules H_2O is dissociated (it means "split") into ions of hydrogen H^+ (i.e., the positively charged H) and of hydroxyl OH^- (negatively charged OH). The concentration of both (H^+ and OH^-) is the same in pure water. Both values equal 10^{-7} eq/l (a unit describing the concentration). Written in more understandable units, the amount of H^+ is 1×10^{-7} g in 1 l and the amount of OH^- is 17×10^{-7} g in 1 l (or 0.0000001 g/l and 0.0000017 g/l). These concentrations are really very small. The value of the water solution reaction does not depend upon the mutual ratio of weights – or more exactly – upon the ratio of masses of H^+ cations and OH^- anions. Instead, the most important conclusion is that everything depends upon the number of positive and negative charges.

When we write pH, we mean the negative value of the logarithm to the base 10 of the H^+ concentration in the liquid. For a value of 10^{-7} eq/l, we obtain log $0.0000001 = -7$ that indicates a neutral reaction of pH=7. Similarly, if we were quantifying the concentration of OH^- for a neutral reaction in pure water, we would write p(OH)=7. When salts of acids or bases are present in soil water, pH values change. If an acid is present in the solution, the H^+ concentration is increased and the value of pH sinks below 7. The value of pH becomes smaller, not larger, with increased acidity because of the negative exponent, i.e., 10^{-5} eq/l is a larger value than 10^{-7} eq/l. Generally, a pH value below 7 means an acid reaction, the presence of acids, and a dominance of H^+. When pH is above 7, then the reaction is alkaline, bases are present in the solution, and it means a dominance of OH^-.

In soils we accept any value of pH within 6.5–7.2 as a characteristic value of a neutral reaction. If pH is above 7.2, the reaction is considered alkaline. If pH is below 6.5, the reaction is acid. Both kinds of reaction are graduated from weak to strong. We should not forget that the pH number is the exponent of 10. Consequently, if we compare soil one having a pH of 6 with soil two having a pH of 7, we must remember that the concentration of H^+ ions in the first soil is ten times higher than that in the second soil. In addition to acid, neutral, and alkaline reactions, we distinguish two other kinds of reactions in soil science, namely, active and exchangeable reactions.

An active reaction in soil science is determined after soil is mixed with neutral water in a ratio 1:2.5. We deal with exchangeable reaction in the next subchapter. In order to provide a comparative scale, we list here a few values of pH in solutions known in our everyday life. The acid in an accumulator or battery for a car has a pH below 1. Our gastric juices have pH = 2.0. Citron juice has pH = 2.4. Vinegar's pH is 2.9. Ceylon tea has pH = 5.5 and milk's pH is 6.5. Our blood has a pH of about 7.4 and seawater about 8.0, and the pH in soap can reach a value up to 10.

Water contained in soil pores is never pure because at each and every moment, it always contains dissolved mineral substances. The weathering of rocks continuously produces simple soluble mineral substances, and even humification culminates with simple dissolved mineral products being added once again into water. Hence, the primary liquid in a soil is actually an aqueous solution of inorganic and organic substances that soil scientists commonly refer to as *soil water*. In other words, soil water is the appropriate name for the liquid phase of a natural soil because it implicitly acknowledges the fact that it is not pure and always contains soluble materials without being specified. Among those diverse, ever-present soluble materials are essential plant nutrients. A logical question would be: Why are they not flushed out of the soil during the infiltration of rainwater? A quick answer: soil does not behave like a strainer. With the great majority of soil pores having sizes well below 1 mm, water is retained in the same manner as if it were in the narrow part of an eyedropper being retained by capillarity. In the next chapter we shall explain more about it, about the forces retaining water in the soil, about the nature of soil water, and the movements of water with and without its solutes inside the soil.

When we summarize all of these soil processes, we find a superficial resemblance with an endless carnival where masks are frequently changed and the selection of dancing partners sometimes appears limitless. With more detailed, comprehensive linkages of observations, we recognize and eventually understand the very strict rules of that soil carnival.

7.5   Specific Surface of the Soil on the Farmer's Field

The walls of soil pores are in contact mainly with liquid soil water, and only rarely is there a direct contact between the solid phase of the pore wall and the gaseous constituents of a soil. Various chemical and biological processes occurring within the liquid soil water influence the solid phase of the entire soil including all of its pore walls. The soil-specific surface and its reactions with neighboring films of soil water play an important role in soil evolution and in relationships between the soil and the flora and fauna that completely depend upon the soil during their life cycles.

Let us imagine a cube having edges that measure 1 cm. If the density of the cube's material is 1 g/cm^3, its surface area is 6 cm^2, and the specific surface related to the mass of its solid phase is 6 cm^2/g $= 0.0006$ m^2/g. Let us further suppose that we have a special saw to cut this cube into smaller pieces without producing any sawdust. Hence, sawing the original cube into cubes having an edge of 0.1 mm produces 10^6 cubic particles each having the surface 0.6 mm^2, and the specific surface of all these small cubes will be 0.06 m^2/g. If we continue to ideally saw each of the cubes into much smaller cubes having an edge of 1 μm, the specific surface of all of these micro-cubes will equal 6 m^2/g. It is therefore entirely logical and not surprising that clay minerals have specific surfaces ranging between m^2/g up to several hundreds of m^2/g. Let us recap some earlier written data. The specific surfaces of clay minerals range between tens and hundreds of m^2/g. Kaolinites have the smallest specific surface – always below 15 m^2/g. And smectites and montmorillonites have the largest specific surfaces, usually slightly below 400 m^2/g. Their large internal surface responsible for such high values is a result of their octahedral layer being sandwiched between two tetrahedral layers. In contrast, owing to the partial binding of these triple layers by K^+ ions, the specific surface of illites is distinctly reduced to values usually between 40 and 90 m^2/g.

The specific surface of a soil depends upon its texture and the type of clay minerals within its clay particles. It usually has a very high value that ranges between 20 and 90 m^2/g. Excluding the consideration of sands, the soil-specific surface supplying essential elements and nutrients to roots of plants growing on a square meter area of an "average" field soil is about 20 km^2. Recognizing that the specific surface of real soils also depends upon the content and quality of humus, its magnitude typically ranges from 10 to 40 km^2 except of sands where it sinks well below 10 km^2. Hence, the first approximation of the specific surface of the root-containing soil layer below an area of 1 m^2 in a farmer's field is 20 km^2. If this value together with its uncertainty owing to spatial variations of soil texture, clay mineral composition, and humus content is extrapolated from 1 m^2 to the size of an average farm in the USA (approximately 440 acres), we obtain the specific surface of the root-containing soil layer that ranges roughly from 20 to 60 million km^2. Although difficult to imagine, we now realize that the specific area of soil particles within the root zone of only one average US farm is even larger than all of the continental area of North America occupied by the USA. Without a doubt, and across all continents, the soil-specific surface is a very important quantitative characteristic having an influential role in many processes decisive for soil fertility, for the role of the soil in hydrologic cycle, for plant and animal life across the landscape, and generally for the quality of the global ecosystem.

We have already explained in Sect. 5.2.2 (*The Finest Minerals*) that a clay mineral surface is never chemically neutral and carries a negative charge due to the substitution of cations inside the tetrahedral and octahedral configurations of its crystal lattice. The negative charges of the crystal surface attract cations present in the soil water residing within soil pores. The broken bonds of the surface of the crystal lattice also contribute to the magnitude of the negative charge by behaving like tentacles that capture cations from the soil water. Humic substances behave similarly. The cations and generally all soluble compounds residing in soil water have their origin in mineral weathering, organic matter humification, and animal activity. Especially in the last 10,000 years as agriculture was practiced and accompanied more and more by industrial and civilized ways of life, virtually all desired features of our comfortable life brought about a lot of wastes and soluble compounds into the environment, notably into soils. As these dissolved materials of diverse chemical composition arrive inside soil pores, the cations attracted and kept on the surfaces of clay minerals and humins are changed and redistributed. Those cations are therefore called exchangeable cations, and the soil capacity to bind them is described as cation exchange capacity, CEC, that is usually measured and reported in milliequivalents per gram of soil, meq/g. The term milliequivalent describes the weight of a substance in milligrams divided by its valence. CEC is closely correlated to the specific surface (m^2/g). Values of CEC for principal groups of clay minerals are given in Sect. 5.2.2. The range of CEC in soil textural classes varies from units to tens of meq/g. Sand has a CEC less than 10 meq/g, sandy loam 10–15 meq/g, loam 15–20 meq/g, and clay loam 20–30 meq/g, and clays have the largest range 20–40 meq/g because they are dominated by specific minerals. The negative charge of humins created primarily by the separation of H^+ from carboxyl or phenolic OH^- causes the CEC of humins to be smaller than that of smectites.

The specific exchangeable cations actually bound to soil particles depend not only upon the percentage of individual compounds in the soil water but also upon the diversity of individual cations. For example, among exchangeable cations, Na^+ has the weakest binding while Fe^{3+} has the strongest binding. The adsorbing strength of other cations increases in the sequence of K^+, NH_4^+, Ca^{2+}, Mg^{2+}, and Al^{3+}. The pH of the soil has a great influence upon the CEC and the relative proportions of individual adsorbed cations. When we measure the CEC of a soil, because we do not usually separate the organic from its inorganic components, we denote all of its sorption-causing materials as a sorption complex.

When the soil is changing toward acidic conditions, the ratio between concentrations of H^+ and OH^- ions is also changing with H^+ concentrations beginning to increase and eventually prevailing. At such times, with a proportion of exchange positions being occupied by H^+, the CEC will not be fully occupied by basic cations such as Ca^{2+}, Mg^{2+}, K^+, and Na^+. The ratio between these adsorbed basic cations and CEC is the base saturation percentage. High values of this base saturation percentage indicate that the soil is a favorable medium for the production of cultural plants. Low values of base saturation indicate that acid conditions and associated processes should be ameliorated by liming accompanied by fertilization of ingredients containing basic cations. Without applying both steps in the sequence, first liming and then fertilization, the acid processes cannot be stopped. Soils in mild climatic regions have a tendency to gradually acidify because they receive a natural supply of free H^+ ions at higher rates than their bases are released by natural processes. The higher the intensity of agriculture, the higher is the "export" of bases by harvesting. Hence, the disproportion between "export and import" of bases increases. In humid tropical regions, the natural processes and outwashing of bases are so intensive that soils are strongly acidic without any human contribution. Extreme values of pH have a major impact upon the availability of important plant nutrients. Because the entry of phosphorus and molybdenum into soil water is strongly reduced under low pH, plants suffer from a shortage of both. Similar situation exists with zinc under high pH. On the other hand, the increased solubility of some elements such as aluminum and manganese under low pH yields toxic conditions for many plants.

It is a relatively rare situation when we precisely measure an exactly neutral soil reaction value of pH = 7. Typically, a neutral soil reaction is denoted whenever an individual measurement of pH falls anywhere between values of 7.2 and 6.5 (see also Sect. 7.4). Traditionally, after we have mixed soil with pure water and measured its pH, we obtain a value designated as its active pH. On the other hand, if we chemically push out (extract) all exchangeable cations (all bases together with the adsorbed H^+) from the soil, a pH measurement of the extract quantifies the soil's exchangeable acidity. The higher the difference between active pH and this "exchangeable pH," the lower is the saturation of CEC by bases and the higher is the need of liming the soil.

7.6   A Carnival with Change of Masks Again?

The change of chemical composition of soil water filling soil pores does not simply mean that the presence of one type of adsorbed cations has been replaced by another type of cation. Some cations have a very thick hydration envelope. For example, Na^+ has an exceptionally large hydration envelope. If these highly, individually hydrated cations are adsorbed on the solid surfaces of fine soil particles, their surrounding envelopes of water do not allow them to get intimately close to the negatively charged surfaces of clay minerals. Inasmuch as water within soil pores is in constant molecular motion regardless of how fast the soil water moves or does not move through the soil, positively charged exchangeable cations remain next to the stable solid surfaces of clay minerals balancing their negative charge. As a result, each solid negatively charged particle is surrounded by its balancing cations and remains unaffected by any water movement. If, however, there is a distinct ratio of monovalent Na^+ present among the exchangeable cations, the solid surface continues to display a certain amount of its negative charge without compensation owing to the great distance of many Na^+ cations away from its surface. The very thick hydration envelopes of Na^+ cause this substantial distance. Consequently, the solid particle surfaces appear negatively charged from the point of view of any moving water. These persisting negatively charged particles do not attract each other. Quite the opposite, they repulse each other. And as a result, they cannot form coagulated units nor is the genesis of microaggregates possible. Owing to this absence of formation of micro-flakes and other types of mutually bound soil clay minerals and silt particles, micro- and macroaggregation do not exist.

If the soil pores in a naturally developed well-aggregated structured soil having both micro- and macroaggregates are penetrated by water containing solutes of Na minerals, the initially present exchangeable bivalent cations within the pores start to be expelled and replaced by Na^+ ions that each surrounded by their own hydration envelopes. Retaining their negative charge, the previously neutral behavior of the soil particles is lost, and they repulse each other while the aggregation disintegrates. The higher is the concentration of Na+ in the solution penetrating into the pores, the stronger is its impact until the original structure is sufficiently transformed to that of a typically behaving structureless soil that becomes slushy and swells when wetted by rain or irrigation water. Its pore-size distribution has been changed substantially with macropores disappearing and microspore distributions displaying a high abundance of very fine pores. When the transformed soil begins to dry, it shrinks and deep, wide cracks appear instead of the earlier original natural macropores, and the abundance of very fine micropores remains or even increases. Soil scientists classify soils according to values of exchangeable sodium percentage, ESP. Because a value of ESP above 10 usually causes already the mentioned problems, soils are denoted as being sodic – or in greater detail, with low, medium, or strong sodicity. However, the sodic problems start at lower ESP if the concentration of salts in the water in soil pores is low. High concentrations of salts in soil water act upon exchangeable Na+ through something like osmotic pressure that presses the cations closer to the solid surface. As a result, the excessive negative charge is reduced, and the disruption of micro- and macroaggregates is less intensive. A sodic soil having a high concentration of salts is called a saline sodic soil. Nevertheless, the actual action of exchangeable Na+ in a saline sodic soil also depends on the composition of all of the salts dissolved in its pore water (Fig. 7.10).

The destruction of soil structure changes the frequency of different sized pores from a distribution of two peaks to only one peak. Earlier, we compared the pore-size distribution curve of a well-aggregated soil to the two humps of a Bactrian camel. Now another similarity becomes apparent – the destruction of soil structure changes the pore-size distribution to a curve having only one peak, comparable to the lonely hump of a dromedary camel. Is not the analogy similar to the change of

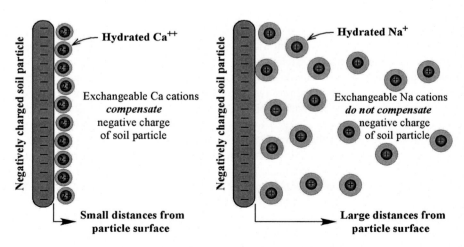

Fig. 7.10 *Left*: exchangeable calcium cations completely compensate the negative charge of a soil particle surface at the shortest distance possible. *Right*: exchangeable sodium cations do not fully compensate the negative charge of the soil particle surface

masks at the carnival and to the long-lasting hangover experienced by the partici-
pants of the carnival?

If the farmers wish to ameliorate sodic soils and prevent the loss of soil structure,
they must introduce a higher concentration of calcium into the soil. The high Ca^{2+}
concentration required to replace the exchangeable Na^+ in exchange positions is
achieved by the application of gypsum. The Ca^{2+} cations in gypsum ($CaSO_4·2H_2O$)
displace and release the Na^+ on exchange positions. The released Na^+ reacts with
sulfate (SO_4^{2-}) to form sodium sulfate (Na_2SO_4), which is a highly water-soluble
compound that is easily leached from the soil. Gypsum is therefore very useful
when soil structure deteriorates because of high sodium.

Up to now we have been focusing on the harmful action of inorganic salts containing sodium. Other inorganic salts containing monovalent cations like NH_4^+ and
to a much lesser extent K^+ also create similar yet somewhat less harmful situations.
The total effects associated with the behavior of these and a few other specific cations frequently observed in arid and semiarid soils of agricultural regions justify the
backbone of ameliorative uses of irrigation. Research leading to the desired control
of specific solutes within such regionally irrigated soils is one of the many primary
goals of soil science.

Nowadays however even in mild climatic regions, intensive agricultural cultivation leads to water pollution, to the salinization of water, or to pollution by other soil
structure-breaking compounds.

7.7   Inhabitants of Soil Pores

Soil pores hold more than just water containing solutes and chemicals and air containing water vapor, CO_2, and other gases. They are also inhabited by various types
of living bodies, many of which are so small that they only can be seen under a
microscope as we have shown in Chap. 4. Living microorganisms make up 60–80 %
of the majority of living flora and fauna that occupy soil pores. The root systems of
plants represent 5–15 %, while macro- and mezzo-organisms account for the
remaining percentages of soil life. The amount of bacteria living in soils is comparable to those occupying water between continents and is substantially higher than
those above the pedosphere of the continents.

Soil bacteria form natural communities influencing the very existence of soil and
its physical properties, the use of nutrients via the root system of plants, and generally the soil fertility, as some species decompose the remnants of animal bodies and
of dead parts of plants. The useful plant nutrients are thus released. Bacteria living
in close proximity to plant roots are called rhizobacteria. The term comes from the
Greek *rhiza* meaning root. Rhizobacteria form symbiotic associations with plants
through their activity in the rhizosphere – the space where roots penetrate the soil.
This relationship is mutually beneficial for the plants and the bacteria. The bacteria
activity in the rhizosphere is 10–100 times more intensive than in soil without
vegetation. Bacteria profit from their existence on and near the roots in many

aspects, primarily because exudates of roots offer nutrients rich in carbon and sugars. Some of the bacteria even enter into the roots and reside within the plant's many vascular bundles without causing any harm to them. Outside in the rhizosphere, bacteria enable prolongation of fine roots. Their production of antibiotics helps the plants in their fight against the harmful microorganisms causing various plant diseases. As the symbiotic bacteria are able to enter into all organs of the plant, they produce special compounds in vascular bundles warning which specific part of the plant is ill. These biochemical messages, equivalent to the recently developed SMS (short message service) available today using cell phones and iPads, have been a part of a plant's life for millions of years. One of the most beneficial of all processes performed by rhizobacteria is nitrogen fixation since nitrogen in gaseous form N_2 is not usable to plants. Nitrogen gas consists of individual molecules of pairs of N atoms bound together by three bonds with sufficient collective strength that plants do not have enough energy to break any of the triple bonds. Fortunately, rhizobacteria are able to convert gaseous N_2 into ammonia NH_3 that is soluble in water and easily metabolized by plants. The enzyme enabling this transformation requires special conditions which are provided by membranes within root nodules. The host plant supplies the bacteria with amino acids so that they have no need to consume ammonia. Nevertheless, the symbiosis between rhizobacteria and plants is not free – it costs each plant about 10–25 % of its total photosynthetic output. Legumes, plants that have root nodules containing rhizobacteria, are recognized as the most efficient of all plants to transform nitrogen from insoluble to soluble forms. The accessibility of other important plant nutrients like iron or phosphates is also frequently enabled by rhizobacteria.

Other microorganisms and zooedaphon inhabiting soil pores were discussed in detail in Chap. 4.

Chapter 8
Soil Is Never Without Water

Regardless of the circumstances, it is undeniable that some amount of water resides within all soil pores across the natural landscape. The amount can be easily determined by merely weighing a small soil sample, drying it at temperature 105 °C for about 6 h, and weighing it again. The difference of the two weights – those before and after drying – is the weight of all water initially present in the soil. When we divide this value by the weight of the oven-dried soil and multiply by 100, we obtain the weight percentage of the soil water content. This value lacks the information needed to ascertain which portion of the soil pores or what fraction of the total porosity was filled by water. To obtain this important information, we have to divide the volume of water expelled from the soil as a result of drying it at 105 °C by the initial volume of the soil sample before it was dried. Inasmuch as the density of soil water is nearly identical to 1, i.e., its weight in grams is numerically equal to its volume in cm^3, we obtain the volumetric fraction of soil water content by dividing the weight of water initially present in the soil by the initial volume of the soil sample. Multiplying this fraction by 100, we obtain the volumetric percentage of soil water content. And from this calculated value, we recognize immediately which portion of the soil pores or what fraction of the total porosity was filled by water.

Practical example: because soil is frequently sampled with cylinders having a volume of 100 cm^3, we pushed one of them into the soil. Upon removing it from the field, we cut away the soil in excess above and below the edges of the cylinder. After weighing it and subtracting the previously measured weight of the empty cylinder, we obtain 164 g – the net weight of the fresh soil sample. After thoroughly drying and weighing the sample within the cylinder and again subtracting the weight of the empty cylinder, we obtain 138 g – the net weight of the dry soil having a volume of 100 cm^3. Subtracting the net weights of the fresh and dried sample (164–138 g), we learn that 26 g of water or a volume of 26 cm^3 was expelled from the fresh soil. Hence, the volumetric soil water content of the fresh soil was 26 %. If the total porosity of the soil was 48 %, then about 54 % of the pore space in the fresh soil was filled by water.

© Springer Science+Business Media Dordrecht 2015
M. Kutílek, D.R. Nielsen, *Soil: The Skin of the Planet Earth*,
DOI 10.1007/978-94-017-9789-4_8

Volumetric soil water content plays many essential roles in our knowledge of hydrological balancing of soil water regimes, of computing the consumption of water by plants, of estimating wind and water erosion of soils, etc. If the volumetric soil water content measured before a brief rain was 35 %, and immediately after the rain a distinctly wetted 10-cm layer had a water content of 47 %, we learn that 12 mm of the rain infiltrated the topsoil. If the rain manifested a measured height of 12 mm, we also learn that all of the rainwater infiltrated the soil surface. However, had the rain been 15 mm, we would know that 3 mm of rainwater was lost by surface runoff or accumulated in puddles made on the soil surface. Various types of observations and deductions obtained from knowledge about volumetric soil water content shall be discussed in subsequent chapters.

Although the above sampling procedure to quantify the amount of water in a soil is very simple and easily understood, it has the serious disadvantage of requiring that holes be dug into the soil surface or the steps of excavated soil using any one of the many types of soil augers. In other words, whenever the soil water content is measured, a hole must be dug. To measure a change in soil water content during a specific period of time, a minimum of two holes are required. The observation of periodic changes of soil water content requires periodic digging of more and more holes. Hence, a long-lasting series of periodic observations destroys the properties of a naturally occurring soil profile and renders it into a material similar to Swiss cheese. The holes after digging and drilling enable preferential deeper flow of rainwater into the soil profile, filling the holes and all excavations with any soil or soil materials does not adequately substitute for the original soil, and the data obtained after almost any period of such observations were not at all realistic and usable. As a result, new methods were developed to measure soil water content without destroying the original nature of soils.

8.1 Methods of Measuring Soil Water Content

Considering the most broadly used present-day procedures, the first method for measuring soil water content without destroying the continuity of a natural field soil was the neutron probe. Later on, dielectric techniques and dual needle heat pulse methods were developed.

Neutron probes were developed about 50 years ago at the time when many radio-isotopes were started to be used in the study of various material properties. The great advantage of the method when compared with the up-to-that-time gravimetric soil sampling was the permanent stable position of the access tube placed into a soil profile to any depth with minimal disturbance of local soil properties. A source of fast neutrons was inserted into the access tube made of either steel or aluminum. Today, the source consists of a mixture of radioactive americium (^{241}Am) and beryllium (Be). Actually, ^{241}Am emits alpha particles that are absorbed by the Be nucleus and emits fast neutrons. Fast neutrons emitted from the beryllium interact primarily with hydrogen atoms of water within the soil. This interaction slows down the fast moving neutrons, and the more intensive is the interaction, the greater is the number and the final concentration of slow neutrons that form a sort of a spherical cloud in the soil. See Fig. 8.1. The radius of the slow neutron cloud usually ranges between 10 and 15 cm, but in extremely dry soils it could be up to 30 cm. Smaller radii prevail in wet loams and clays, while larger radii develop in dry soils of all textures. The concentration of slow neutrons is measured by a special counter.

A fast neutron is slowed down the most when it collides with the proton of a hydrogen nucleus. The principle of the strong conversion of fast neutrons into thermal neutrons depends upon the similarity between a neutron and hydrogen – their sizes and masses are virtually identical. When a tennis ball hits the hard surface of a tennis court, it bounces and continues to move with a just slightly reduced velocity. However, if the ball hits another ball lying quietly on the court surface, the flying ball loses its original velocity giving roughly half of its energy to the ball initially at rest. Frequently, both balls finally roll on the surface approximately in the direction of the flying ball before the collision. Or during tennis practice, after a tennis ball hits the wall, it comes back to the practicing tennis player with only a slight reduction of its velocity. The change of the velocity of the ball after its collision with another object depends upon the ratio of the mass of the ball to the mass of the object. Our observation of the collision of flying ball with another initially at rest is a rough model of what happens if a fast neutron hits a hydrogen nucleus. In reality a very fast neutron must collide many times with many hydrogen nuclei in order to slow down. After many such collisions, the neutron approaches the same velocity of other neutrons that is consistent with the prevailing temperature of the soil. Hence,

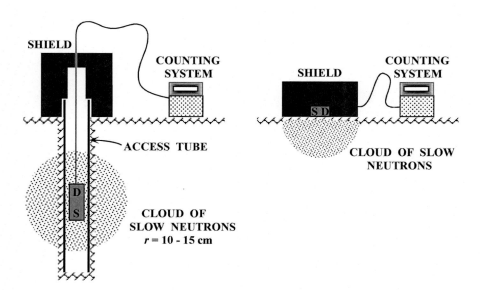

Fig. 8.1 Schemes of neutron probe for measurement of soil water content in the soil profile (*left side* of the figure) and on the soil surface (*right side*). The source of fast neutrons is denoted by S and the detector of slow neutrons by D

we speak about the thermalization of fast neutrons and about thermalized or thermal neutrons. In order to measure the concentration of these thermalized neutrons, a special detector of slow neutrons has to be placed immediately next to the source of fast neutrons. Pulses of slow neutrons are preamplified and sent to the counting system. Generally, the higher is the content of thermalized neutrons, the higher is the content of soil water. A calibration between counted pulses and volumetric soil water content is developed. Electronic drifts, temperature migrations, and counting time variations are avoided by dividing each count rate by that of a reference material placed inside of the protective shield. A typical standardized calibration curve manifesting a straight line between count ratio and volumetric soil water content is illustrated in Fig. 8.2.

Because various amounts of hydrogen also occur in different clay minerals and in different amounts and kinds of humus, each soil has therefore its own specific calibration requirement. A typical average linear calibration can be adjusted for each specific soil or field location by measuring paired values of count ratio and volumetric water content when the soil is relatively dry and wet during rainless and rainy periods, respectively. Dealing with soils in arid regions, we can wet the soil by ponding water on a small surface area of a few square meters bounded by low dikes.

The advantage of the neutron method follows from one simple example. We placed two access tubes in two neighboring small fields with the same soil. Field number one was plowed, field number two was not plowed, and both were without vegetation. The fields are on a large plain having groundwater level at a depth of

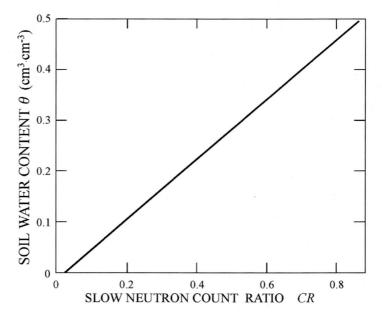

Fig. 8.2 Example of calibration curve of soil water content θ (cm^3·cm^{-3}) versus count ratio CR of slow neutrons

4 m. On the 15th and 25th of May, we measured the soil water content with a neutron probe. This 10-day period was without rain. When we found a difference in water content, it was due to the difference in the evaporation rate of water from the soil, and we simply measured how much water evaporated. If there was a difference between the two fields, it was due to the tillage and no tillage. Research running for a long time during the growing season to study plant growth variations can easily be done without repetitive drilling, frequent soil sample removal, and disturbing the soil and its properties. And with the use of only one access tube, comprehensive studies of the infiltration, evaporation, and redistribution of rainwater are feasible for both uniform and layered soils. Using more access tubes for large areas together with spatial and temporal statistics, water balance studies can be linked to the causes of specific soil-water-plant relationships without establishing traditionally replicated plots with different levels of arbitrary treatments.

It is worth mentioning that the start of the neutron method was not easy. The application of neutron scattering effect to measure soil moisture in field soils was developed in 1950 by John (Jack) F. Stone. He hand made a portable neutron meter before such moisture meters were commercially available. He did so by following up on the research of Wilford Gardner under the leadership of Don Kirkham, a professor at Iowa State University. Jack's meter was powered by several kinds of dry-cell batteries. At that time just after Jack had successfully tested his homemade meter, Kirkham gave a seminar in the Physics Department on the application of the neutron scattering to measure soil moisture. I (DRN) attended his seminar in the Physics Department when all of the attending theoretical physicists criticized and even laughed during the discussion of his presentation. I remember well Kirkham's calm reaction and especially his final remark at the end of the seminar – "I appreciate your comments that such a method is theoretically impossible, but Jack's field measurements and those in large soil containers prove that neutron scattering can be used in a very practical way to measure soil water content." It was exciting for me to interact with Jack in 1955–1956 while he tested the new instrument that was soon commercially produced.

Lastly, but nevertheless essential and necessary, we mention that operators of a neutron probe must be protected by a shield against the harmful radiation.

Time domain reflectivity (TDR) is based upon measurements of dielectric permittivity ε. Permittivity is a characteristic that relates to the ability of a material to transmit or "permit" an electric field. Although it is sort of similar to electrical conductivity, it is a measure of the charge separation, not the current, when a voltage is applied under a specific condition. The charge that moves within the material is bound to the positive charge of the nucleus and to the negative charge of the outside electrons. Figure 8.3 illustrates a very simple model of what happens to molecules within a material subjected to an electric field. Since ε of water is about 80, that of air is nearly 1, and that of soil particles ranges from 3 to 5 in sands and from 8 to 20 in clays, a measured value of ε for any soil is closely linked to its water content provided that other conditions are not changed. This provision is fulfilled since the soil air humidity is always high and the composition of the soil solution remains nearly constant.

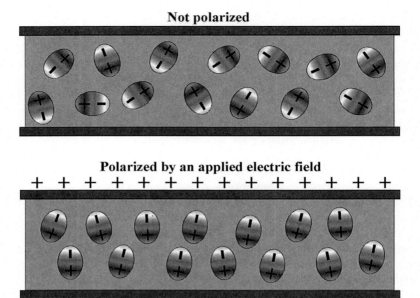

Fig. 8.3 Polarization of water molecules due to the applied electric field changes the dielectric permittivity ε of the porous material where the pores are partly or fully filled by water. The value of ε depends then upon the extent of filling the soil pores by water and this is the principle of TDR (time domain reflectivity) method

Classical TDR instrumentation consists of four basic units: a timing circuit, pulse generator, sampler, and display. The pulse generator transmits electromagnetic waves along a transmission line leading into the soil. One of the commercially available instruments has three rods at mutually fixed positions that are usually permanently inserted into the soil or, in some instances, pushed into the soil at the time of measurement. Other instruments rely on access tubes for their insertion into the soil at a desired depth. Changes in the soil impedance generated by the electromagnetic waves are reflected, signaled back, and recorded. By measuring the time required for a sequence of pulses to travel along the known probe length, the dielectric constant of the soil can be computed and used to obtain the volumetric soil water content. Calibration curves must be made for each soil taking into account its specific porosity or degree of compaction.

Numerous other methods have been developed to measure the amount of water in soil. Here we mention only a couple of examples. One is based on the adsorption and scattering of gamma radiation. Because of strict, necessary geometric conditions imposed on the radiation and its measurement, the method is more applicable to soil columns measured in a laboratory than to natural soil profiles. In the field, the soil water content is adequately estimated between two access tubes only when they have been installed exactly parallel to each other. The other method utilizes spectral analysis of Earth's visible and infrared radiation for estimating soil water content within and across large land areas. As a remote sensing technique, the radiation is detected and measured either from a balloon or, more frequently, from an airplane or satellite.

8.2 How Water Is Kept in Soil Pores

When it rains on a rock, and if it is a heavy rain, little or no water is retained and it flows away except from depressions in the rock surface where small isolated water puddles could appear. On the other hand, the majority of rains change their appearance on the rock surface from individual drops to a continuous surface flow just at the beginning of rainfall and briefly after it.

When it rains on a soil surface, either all of the water or at least a portion of it infiltrates and is captured and held within soil pores. Why does the infiltrated water not flow through the soil to the depths where it would not be of any use for plants? Except within a few centimeters from its surface, why does the soil remain wet even for a couple of weeks after the rain? Retained rainwater evaporates first from the surface, while the soil below the thin surface layer remains nearly at the same water content as it was when the rain stopped. We recognize it when we remove a lump of soil from below the dried-out surface layer and subsequently squeeze it in the palm of our hand. Opening our palm, we observe that the skin of our palm is very wet and perhaps even partially covered by small drops of water. With such an experience we realize that some of the rainwater was kept in the soil while other portions of it were pushed out. The mechanism is similar to what we know from our household activities. After we take a wet sponge from a pan of water, squeeze it with all of our might, and wipe a table, the squeezed sponge leaves a wet trail across the surface of the table. It is similar with soil. Although we pushed a small amount of water into our palm by squeezing the lump of soil, some water remained strongly bound within its fine soil pores. Water is kept in fine pores mainly by *capillary forces*.

In reality, soil water is never immobile. It continually wanders, migrates, and flows at different rates from one location to another. But in some instances its instantaneous flow rate is so slow that we can simplify the situation with a hydrostatic model where all forces acting upon water are in equilibrium and where only the laws of hydrostatics are valid. After discussing wetting angles between water and solid surfaces, we shall explain the principles of capillarity and those of adsorption.

When water meets or contacts a solid surface, it has three potential impacts: (a) to completely wet the surface, (b) to incompletely wet the surface, and (c) not to wet the surface. An example of such behavior is shown in the top half of Fig. 8.4 when water rests on top of three horizontal solid plates forming different contact angles γ between the water-solid interface and the water surface. If the solid surface is completely wet, the interface and the water surface coincide and manifest a value $\gamma = 0°$. When the solid surface is incompletely wetted, values of γ range between 0° and 90°. A non-wetting surface is indicated when values of γ are greater than 90°.

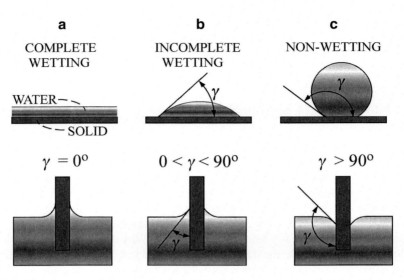

Fig. 8.4 Contact angle γ between water and solid: (**a**) complete wetting; (**b**) incomplete wetting; (**c**) non-wetting

Another manifestation of the impact of a wetting angle is apparent in the bottom half of Fig. 8.4 when we submerge one end of the solid plate vertically into water to observe the shape of water level formed near the plate. When water wets the plate, its surface forms an upward oriented arc as if it were climbing up to the plate. And, whenever the water does not wet the plate, its surface forms a downward oriented arc as if it were stepping down to the plate. When the contact angle is zero, its arc length is one-fourth of a cylindrical circumference having one end in contact with the plain water level and the other end in contact with the vertical solid plate. If wetting is not complete, the contact angle is somewhere between 0 and 90°. Such wetting behavior is known as hydrophilicity and the solid is characterized as *hydrophilic* – derived from Greek *hydros* for water and *philia* for friendship. The lack of wetting of the solid surface by water demonstrated by contact angles greater than 90° is known as hydrophobicity and the solid is described as hydrophobic from Greek *phobos* for fear. It would be a false assumption that hydrophobicity does not exist in soils. It happens when soil particles are covered by thin films of humic substances or whenever humifying residuals of organic bodies cover the soil topographical surface.

Let us now submerge one end of an empty glass capillary tube vertically into water and observe the shape and height of the level of water sucked into it. As water rises, its curved water surface forms the shape of a cup or a hemisphere. We describe this type of "no flat plain" by the term *concave* that means a surface like the inside of a sphere. This curving of the water surface brings our attention to the physical phenomenon of "capillarity" when water molecules attract each of their neighboring water molecules to form and pull together in an organized manner called "cohesion."

Simplifying molecular conditions in the vicinity of a horizontal, flat plain of water, we arbitrarily designate in Fig. 8.5 any one of the uppermost water molecules along the plain with a script *M*. Such a molecule is attracted to molecules located below and horizontally around it. Horizontally, the attraction from a molecule on its right side is annihilated by that from its left side, and the same happens to all molecules on the surface. Hence, horizontally placed molecules have no effect upon our designated molecule *M* at the water level. Because there are no water molecules above our molecule *M*, the top half of it is not upwardly attracted. Cohesion is therefore restricted to the bottom part of our molecule *M*. The resultant of all acting forces is directed downward into the liquid and is described as surface tension in our macro world. Thus, surface tension is caused by the cohesive intermolecular forces of water acting on water molecules at the surface of liquid water.

The language of physics says that surface tension is a force related to a unit length, Newton per meter, N/m. Usually, it is more practical and convenient to use the surface pressure expressed as Pascals, Pa, where 1 Pa is one Newton per one square m, $1 \, Pa = 1 \, N/m^2$. Since it is extremely small unit, we indicate the surface pressure in hectopascals, $1 \, hPa = 100 \, Pa$, or in kilopascals, $1 \, kPa = 1{,}000 \, Pa$. When we measure dried-out soils, we use units in megapascals, $MPa = 10^6 \, Pa = 10^4 \, hPa$. The meteorologists are used to pressure units in millibars, where $1 \, mbar = 1 \, hPa$. The conversion of units is according to the following relations: $1 \, Pa = 1 \, N/m^2 = 10^{-5}$ bar being the equivalent of $0.00001 \, bar = 10.197 \times 10^{-6}$ at (technical atmosphere) $= 9.8692 \times 10^{-6}$ atm (atmosphere) $= 7.5 \times 10^{-3}$ Torr (Torr $= 1$ mm of Hg) $= 1.45 \times 10^{-4}$ psi (pounds per square inch). Surface tension on the flat plain water level is the resultant of cohesion action of the water molecules located below the water level. It is 7.28×10^{-2} N/m or 72.8 dyn/cm at the temperature 20 °C.

Fig. 8.5 Simplified model of the intermolecular forces acting upon the water molecule on the plain horizontal water surface and upon water molecule in the interior of water. The molecule on the surface, attracted in one-sided direction downward into the liquid, is observed as surface tension on the macroscale

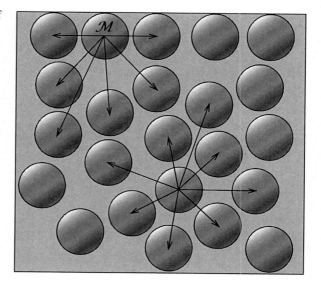

Adhesion, known as the attraction between unlike molecules, potentially occurs when a water molecule comes near a molecule of a solid surface. If this adhesion has a higher value than the cohesion between water molecules, the nearby water molecules look like they are climbing up the vertical hydrophilic surface we observed earlier in the bottom half of Fig. 8.4a, b. On the other hand, when the adhesion between water and solid surface molecules is smaller than the cohesion between water molecules, the water molecules near the hydrophobic solid surface are repulsed and the water forms a convex shape previously seen in Fig 8.4c with contact angle above 90°. A simple observable example is a droplet of water forming a spherical shape in order to minimize contact with a hydrophobic leaf.

Let us now continue our earlier discussion of capillarity when water begins to move upward into a vertical glass tube after its lower end has been submerged in water. On our oversimplified molecular scale, we observe hydrophilic curving where each water molecule of the curved surface is attracted in addition to molecules placed not only on the same level and below but also to neighboring molecules that have "climbed" higher up onto the wall to elevations above the free water horizontal plane surface. Inside the tube the shape of the water surface is "U-like," i.e., concave. The resulting surface pressure of this "U-like" or "cup-like" water surface illustrated by an arrow labeled p_s in Fig. 8.6 is smaller than that of the reference pressure p_r of the horizontal plane water because there are water molecules in the capillary tube above the horizontal water plane that compensate the attractive forces

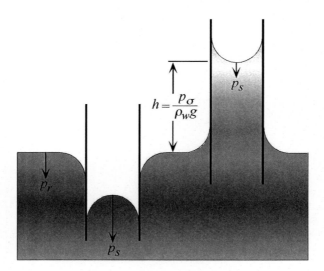

Fig. 8.6 On the *left side* of the figure, capillary depression occurs when water is not wetting the solid walls of the tube. The length of the *arrow* represents the increased surface pressure p_s due to the convex shape of the water level, and this increase pushes down the curved water level in the tube compared to the flat plane water level with surface pressure p_r. On the *right side* of the figure when the water molecules are wetting the solid walls of the tube, capillary rise occurs since the concave shape causes decreased surface pressure compared to flat plain water level. This difference of surface pressures causes the water to be sucked up into the capillary tube

from molecules located below plane. Since the surface tension inside of the capillary is smaller than that away from the tube in the horizontal flat plane of water, water is sucked into the tube until an equilibrium is reached. At equilibrium, the weight of water sucked up into the capillary equals the difference of surface water tensions between the curved and flat plane water levels. The smaller the radius of the capillary tube, the greater is the mentioned difference in surface water tensions and the higher is the column of water inside of the capillary, provided that the contact angle does not change, i.e., the hydrophilicity does not change.

The simplest model of water in soil is composed of parallel vertical hydrophilic capillary tubes with radii from the smallest tube on the left side increasing to the largest on the right side of the model as shown in Fig. 8.7. Owing to the hydrophilicity of the capillaries, after inserting the bottom end of this model into free water, the water level inside each tube rises to the height related to its particular radius. Hence, the highest water meniscus occurs in the smallest tube on the left hand side, and as the radii gradually increase, the water menisci gradually reach their lowest height within the largest tube on the right side. Connecting each of the heights of their menisci, we obtain a smooth curve. A similar curve is obtained when the heights above the groundwater level are plotted against field-measured soil water content values of a silty or fine sandy soil.

Comparing this field-measured curve with that illustrated in Fig. 8.7 derived from the capillary model leads us to the conclusion that for some simple cases, the behavior of soil water is successfully mimicked by the parallel capillary tube model. The success of the model was judged on the basis of soil water content measurements

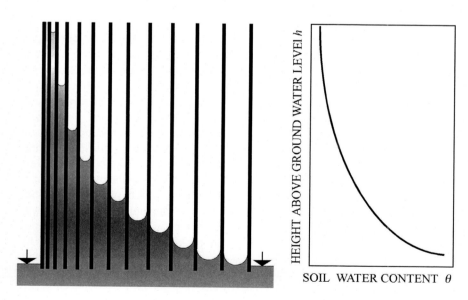

Fig. 8.7 On the *left*, a simplified capillary model illustrates the soil water content above groundwater level. On the *right*, after plotting measured data we obtain the smooth curve of water content distribution above groundwater level

taken at small depth intervals just above a constant water table. Such samples may be obtained from a pit dug to the depth of a water table that occurs at a relatively shallow depth close to the soil surface. The flow of water from the undisturbed soil eventually reestablishes a water table at the bottom of the pit. After waiting several more hours, there is an equilibrium between the height of the capillary rise of water within the soil pores and the distance above the free water level in the pit.

When the lower end of a vertical capillary tube made of *hydrophobic* walls is submerged into water to a sufficient depth, the water moves upward into it to a level below that of the horizontal free water surface and manifests a curved upper surface in the form of a spherical cap, i.e., convex. Indeed, a capillary depression is created because water molecules along the spherical cap surface are attracted to higher numbers of neighboring water molecules than those on the free water surface. The resulting surface pressure of the spherical cap water surface illustrated by an arrow labeled p_s in Fig. 8.6. is larger than that of the reference pressure p_r of the horizontal plane of free water. This bigger surface pressure does not allow the water to enter into the height of water level in hydrophilic tube. The difference of surface pressure $p_s - p_r$ is the pressure by which the water level is pressed down, when compared to horizontal plane of water level. Humins may have hydrophobic properties and if they form a film partly covering mineral particles, the soil could be in some instances partly hydrophobic, especially if it is dried out. When it rains after a dry period, we can observe that the first raindrops form small spheres that roll across on the dry soil surface. This effect of hydrophobicity of an excessively dry soil surface rich in humic substances usually disappears soon after it continues to rain to moisten the topsoil.

Up to now we only considered capillary forces being responsible for keeping water within a soil. However, there are many other forces continually acting upon soil water that should not be ignored. Hence, we next speak and make a simple experiment about adsorptive forces. Taking a clod of loam soil, we crumble it, lay it out on a plate in a thin layer of about 1 cm, and keep it at room temperature until it looks completely dry. At that time we take a 10-g sample of the apparently dry soil, place it in a hot oven for several hours, and weigh it again. We set the temperature of the oven at 105 °C – high enough to remove water molecules adsorbed on soil particle surfaces yet not sufficiently hot to destroy OH^- bonds within soil constituents. To our surprise when we remove the sample from the oven, its weight is less than 10 g. The difference in its weight before and after being in the oven is the mass of water removed from the soil. Let us suppose that the weight of the sample after drying was 9.5 g. We released 0.5 g of water from the apparently dry soil that was bound mainly by adsorptive forces and not by capillarity. The water content of our apparently dry loam used in our illustrative example was 0.5/9.5 = 0.053 or 5.3 % by weight.

Genuine adsorption of water readily occurs in completely dry soils when they are in contact with humid air from which water molecules are attracted to the solid soil particle surfaces. When the air humidity is about 20–30 %, a continuous layer having an average thickness of one water molecule is completed. Although such very dry air occurs mainly in desert regions under natural conditions, it could be purposely established in the laboratory to use the equilibrated soil water content as the basis for computing the specific soil particle surface. When the air humidity is increased up to about 60 %, a value occurring frequently in the mild climatic zone, the water layer covering the solid soil surfaces is a film having a thickness of 4–6 water molecules. The thickness of these films is the result of the combined simultaneous forces of both adsorption and capillary condensation. Computed values of the pressure potential of these films range from about 100 to 150 MPa. When the air humidity rises to values in the vicinity of 100 %, water continues to accumulate in soil pores, but the mechanism is only that of capillary condensation with the pressure potential reaching just to about 5 MPa and less.

Mitscherlich, a German soil scientist, proposed in 1901 that the hydroscopic properties of a soil be described and quantified by a term named the hydroscopic coefficient that equaled the soil water content in equilibrium with 95 % air humidity verified with laboratory equipment. Because two phenomena, adsorption and capillary condensation, contribute to the magnitude of the proposed term, its magnitude does not actually isolate, characterize, nor quantify the true hydroscopic property of a soil. Hence, after more than a century, the term with its method of approximation used for many decades is now obsolete.

There is another consequence of the curved water level. The lower is the surface pressure, the less water molecules escape into the atmosphere above the curved water level. Or, in other words, the narrower is the cylindrical capillary and the greater is the curvature of water level, the lower is the partial pressure of water vapor in the atmosphere above the water level; see Fig. 8.8. We model the mentioned

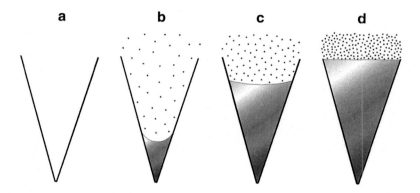

Fig. 8.8 The greater the curvature of a liquid water surface, the fewer are the number of neighboring water vapor molecules – or more precisely, the greater the curvature, the lower is the partial pressure of water vapor in the air next to a water surface. Here, we illustrate this relationship by the density of dots (water molecules in the air) and the amount of water within a triangle-like soil pore: (**a**) No water molecules (*dots*) are present within the completely dried-out soil pore (*triangle*). (**b**) Within the triangle, a very small amount of liquid water resulting from capillary condensation manifests a strongly curved surface in the vicinity of a very low air humidity (a small number of *dots*). (**c**) Capillarity has caused a low curvature of the liquid water surface and a high number of water molecules in the air, i.e., a high air humidity. (**d**) The triangular pore completely filled with liquid water manifests a flat water plain covered by a maximum possible number of water molecules in the air having a humidity of 100 %

relationships on one conical pore. It is drawn in the cross section as a triangle. When the pore is fully separated from the outside environment and when there is no water, the air is absolutely dry. Then we allow air with a humidity of about 20–30 % to enter the pore at a constant temperature, let us say at 20 °C. The individual water molecules are attracted to the solid surface of the pore forming a film of the average thickness of one molecule as a direct result of water vapor adsorption. We continue in increasing the air humidity up to about 50–60 % and the adsorption continues; the film thickness of adsorbed water makes first two, then three molecules. Now, the film starts to form a curved horn at the contact of films and the curvature is dependent upon the air humidity. The process is called capillary condensation and the situation is depicted in B section of Fig. 8.8.

The higher is air humidity, the smaller is the curvature in the horn. Now we add a little bit of liquid water. The meniscus still exists but it is less curved, the surface pressure is increased, and the air humidity has increased, too; see the situation in case C. We continue in adding liquid water until the water level reaches the top of the cone where the capillarity does not exist anymore. The water level is a flat plane. The content of the number of water molecules as vapor in the air is the maximum possible at the given temperature. Since the air is fully saturated by water vapor, the air humidity is 100 %.

However, we have not reached the end of our description of the various kinds of forces acting upon soil water. The air within air bubbles completely enclosed and surrounded by the continuous phase of water within the soil cannot readily escape from the soil. When this enclosed air is compressed by infiltrating water or the weight of agricultural machinery, a soil hydrologist must consider this mechanical component of force acting on the soil water. Moreover, whenever the content and kinds of soluble salts within a soil fluctuates, changes of osmotic pressure also complicate our simple way of studying force fields acting upon soil water. The consideration of osmotic behavior within soil profiles is especially important in arid and semiarid zones where salinity dynamics frequently prevail and shifts of salt content are ubiquitous.

8.3 Does the Soil Like Its Water Every Time?

The friendship between soil and water, i.e., the degree of soil hydrophilicity, differs according to each kind of soil as well as to the predisposing conditions at any given time. If a soil is first saturated by water, then partly dried, then wetted again, etc., each change of soil water content regardless of its frequency alters the friendship or the degree of hydrophilicity. The simplest demonstration of hydrophilicity is using a capillary tube to observe an effect called capillary *hysteresis*. Let us assume that we have two capillary tubes of the same material and of the same radius. We insert the bottom end of one capillary in Fig. 8.9a into water and allow the water to rise due to capillary forces. Next, after totally submerging the other capillary in water, we subsequently pull it partly out keeping its bottom end still below the free water

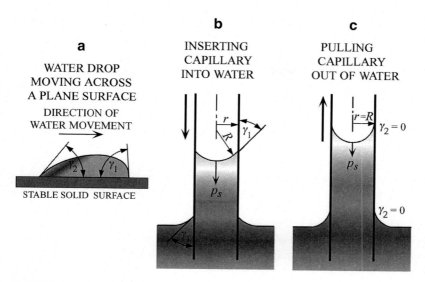

Fig. 8.9 Principle of hysteresis: (**a**) waterdrop moves along the solid surface; (**b**) originally dry capillary tube is inserted into water; (**c**) capillary tube was first fully immersed in water and then partially pulled out. The length of the *arrow* below the curved water level represents the surface pressure, which is different just due to the change of contact angle γ and thus the changed curvature

level (Fig. 8.9b). The wetting angle in the first capillary is greater than that in the second capillary because it is more difficult for water to simultaneously wet and move along an initially dry surface. We note that the curvature of the meniscus of the first capillary is flatter than that in the second capillary. Within the second capillary that was initially submerged in water, water molecules had more time to partially displace foreign molecules and more completely wet its cylindrical wall. The difference in curvatures of the two menisci illustrates that the surface tension is smaller in the second capillary than that pushed only into the water. The surface pressures of the menisci p_s represented by the lengths of arrows account for the height of capillary rise in the first capillary being smaller than that in the second capillary. To be more comprehensive, we have to keep in mind that the size of a capillary tube is not the single decisive factor on the retention of water in soil and that the extent of hydrophilicity (or hydrophobicity) has to be considered.

8.4   From Capillary Tubes to Real Soil Pores Partly Filled by Water

It is obvious that real soil pores have shapes distinctly different from bundles of parallel capillary tubes. Nevertheless, we discussed parallel tube models just for their advantage in simplicity and clear physical demonstration. We have mentioned already in Chap. 7 that natural-occurring soil pores are variously shaped and curved, change in size, occupy irregular spaces between solid surfaces of particles and those within and between aggregates, and manifest a diversity of simple to complex spatial interconnectivity. Because capillarity is linked to changes of water surface curvatures within soil pores on a microscale and consequently to changes of surface tension or pressure, it is important to detect and understand the relationship between soil water content and macroscopically measured capillary pressure. The smaller is the soil water content, the lower should be the pressure at equilibrium. Because this meniscus pressure is negative compared to the pressure on a horizontal, plane water surface, its absolute value is higher. Let us explain with a simple example. We measured the soil water content of a loam to be 32 % when the meniscus pressure was −15 kPa, and when the pressure was −100 kPa, we measured a smaller water content of 26 %. The negative sign indicates that we have to use less energy for extracting a small drop of water from soil having a water content of 32 % than the energy necessary for extracting the same small amount of water from the loam having a smaller water content of 26 %. Or, vice versa, water is bound in this loam by a higher energy if its water content at 26 % is less than that at 32 % or higher soil water contents. Although these values of pressure are often carelessly referred to as the potential of soil water, it is more precise to identify them as pressure potential.

Values of soil water pressure potential should be measured on "undisturbed" soil. Such samples are obtained by pressing a metallic cylinder (volume of at least 100 cm^3) into the soil without disturbing the soil's natural arrangement inside of the ring. The cylinders are extracted together with the undisturbed soil and placed on the porous ceramic plate in the chamber of a pressure plate apparatus; see Fig. 8.10.

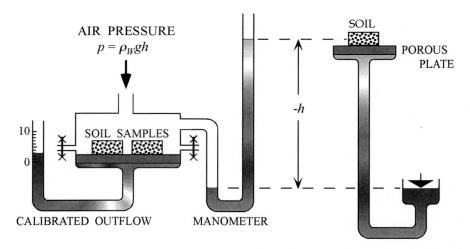

AIR PRESSURE
$p = \rho_W g h$

SOIL

POROUS PLATE

-h

SOIL SAMPLES

10

0

CALIBRATED OUTFLOW   MANOMETER

Fig. 8.10 Pressure plate apparatus on the *left side* is provided with a liquid manometer just for simplicity. Routinely it is connected to a gauge. The applied air pressure pushes out of the soil sample all water kept by capillary pressure (in absolute value) lower than is the applied overpressure, indicated here by water manometer. Its value is the same as in the equipment on the *right-hand side* which practically imitates the decreased water content due to the depth of groundwater level. The apparatus name is tension plate apparatus

The soil samples including the porous plate are initially water saturated. Subsequently, we measure the volumes of outflow water collected in the calibrated tube, or burette, as we incrementally increase the air pressure in the pressure chamber. We start with a very small overpressure, e.g., 2 kPa above the atmospheric pressure in the laboratory, and register the volume of outflow. Next, we increase the overpressure to 5 kPa and again register the volume of water collected in the burette. We continue extracting water from the soil using progressively larger, nonlinear steps of overpressure up to a magnitude of about 2 MPa.

The incremental volumes of water measured in the outflow burette are converted into values of soil water content that were in equilibrium with each applied overpressure. These pairs of data plotted in a graph yield the main drainage curve shown in Fig. 8.11. The shape of this curve, also known as a water retention curve, is roughly similar to that of the curve in Fig. 8.7 derived for the capillary rise model of a bundle of various-sized capillary tubes. The impact of hysteresis introduced in Sect. 8.3 can also be seen in Fig. 8.11 by observing that when water wets an initially dry soil, its water content is less than that of a soil draining from an initially water-saturated content. The experienced soil physicist measures soil water retention functions to derive many important soil properties, e.g., the size of soil pores and their distribution in the soil, the extent of soil compression, the relative ease that soil conducts water, and the relative difficulty for plants to suck soil water into their roots.

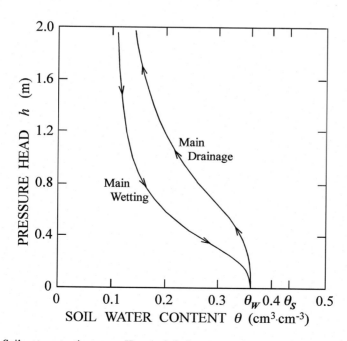

Fig. 8.11 Soil water retention curve. Hysteresis is demonstrated by its two main branches

Chapter 9
How Water Flows in Soil

9.1 The First Observations

From the very beginning of their studies, soil scientists noticed two characteristic values of soil water content. One was when water moving down from the nearly saturated topsoil starts to drastically slow down even though the soil contains still a substantial amount of stored water. If evaporation was prevented, the change of rate of drainage was reflected by the soil water content, which did not appear to change for several days after the initial decrease. Today, it is a common practice to observe such a characteristic by first surrounding a small, flat soil surface of about 1 m² with a small dike having a height of about 10 cm. Water is next steadily ponded on the area isolated from its surroundings until we guess that the soil is water saturated to the depth we wish to study. With the surface no longer ponded with water and covered by a folio in order to prevent evaporation, a limited portion of the soil water flows quickly and readily from the topsoil into the subsoil. The soil water content in the top layers decreases, while the deeper layers are gaining water. In a few days, the soil water content in the topsoil looks as though it is not changing – indeed the downward flux is ten to hundred times smaller than it was when water first started to drain immediately after its surface was not ponded with water. This soil water characteristic was originally called *capillary capacity* and later renamed to *field capacity*. It can be estimated using undisturbed soil samples within metallic cylinders of 100 cm³ volume maintained at approximately 10^4 Pa negative pressure = 100 cm of water column. A rough value of capillary capacity could be read from the soil water retention curve (Fig. 8.11).

The second characteristic value was when the roots of a plant were unable to extract enough soil water for the plant to survive and function – a condition known as permanent wilting that prevented the revival of the plant. At the end of the nineteenth century and during the first part of the twentieth century, it was apparent to all soil scientists that each soil type as well as each textural class was easily associated with specific values for the abovementioned two characteristics. Therefore,

© Springer Science+Business Media Dordrecht 2015
M. Kutílek, D.R. Nielsen, *Soil: The Skin of the Planet Earth*,
DOI 10.1007/978-94-017-9789-4_9

many approximate procedures based on simple, frequently measured data were proposed. Examples of two such procedures are the linear correlations between soil
texture and either capillary capacity or wilting point.

During the early empirical period of soil physics, the flow of soil water was characterized by three phases. The soil flow rate was high and the soil rapidly drained
when the soil water content ranged between saturation and capillary capacity. When
the soil water content ranged between capillary capacity and wilting point, the flow
rate was low. And when the soil water content was below the wilting point, the flow
rate was so extremely low that soil scientists assumed that soil water was not flowing. Nowadays, scientific segments of classical Newtonian physics, hydraulics, and
mechanics are combined to understand and predict the basic dynamic processes
controlling soil water behavior.

9.2   Flow of Water in Saturated and Unsaturated Soils

When water flows on the Earth's surface, people often comment that it flows downhill or downslope. However, this remark is not exact. If it were indeed correct, water
could not flow in a river having a mound or bump across its bottom with a slope in
the opposite direction of the general slope of the riverbed. A more exact remark
would say that water flows in the direction of the slope of the water level of the river.
Even this latter selection of words is still only approximate because below the water
level of a river we find transverse as well as reverse fluxes. Transverse river fluxes
are roughly perpendicular to the direction of the main river flow. Reverse fluxes
commonly exist below a weir with water moving back along the bottom without
regard to the direction of the local bottom slope. The local gradient of the potential
always plays the decisive role in the local fluxes. This is why we will speak about
potential and its slope (gradient) when we are dealing with flow within the pore
systems of soils. In order to accentuate the important role of potential upon the flow
on another much bigger scale, we describe a well-known ocean stream. Bulky and
intense streams caused by differences in potential are ubiquitous in oceans. Let us
start with easily observed and understood causative forces in order to identify the
nature of their active potential gradients.

Essentially as a continuation of the Equatorial Currents, the well-known Gulf
Stream in the Atlantic Ocean starts in the tropical area of the Mexican Gulf. We demonstrate its size by comparing it to the Amazon River. With the narrowest portion of
the Gulf Stream representing about 20 Amazons, its biggest part grows and reaches a
size of 200 Amazons. After circulating in the warm waters of the Gulf of Mexico, the
Gulf Stream exits through the Straits of Florida, flows north parallel to the East Coast
of the USA, enters the deep Atlantic Ocean after passing Cape Hatteras, and brings
relatively warm water to the west coast of Europe. As it continues to flow further
northward, it gives us the impression that it disappears in the Arctic zone. But it does
not actually disappear – it simply dives to the depths of the ocean bottom. This sudden change from horizontal to nearly vertical downward streaming is simply caused
by the gradual cooling of its water to temperatures close to 4 °C. At these temperatures,

the water being more dense and heavier than the surrounding ocean water has the tendency to "fall down," i.e., to stream vertically down to the ocean bottom. At depths near the bottom, the Gulf Stream merely changes its direction and flows to the south even though there is no slope of water level related to its new direction of flow. A similar occurrence of ocean water flowing downward is observed when the salinity of water increases and provides a second reason why the Gulf Stream changes its direction. As the warmer water of the Gulf Stream compared with that of the surrounding ocean moves to the north, evaporation is higher and none of its salts evaporate with water molecules. During its long trip from the Equatorial region to the Arctic Ocean, the salinity of the Gulf Stream increases. Prevailing winds blowing to the East also influence the direction of the Gulf Stream (Fig. 9.1).

Knowing that the flow of water in the ocean on a megascale is influenced by several acting factors, or forces, we shall merely add them together and analytically transform them into potentials. The difference of potentials between two locations, or better expressed as a potential gradient, determines the direction of flow. Moreover, the higher magnitude of this gradient, the greater is the velocity or the rate of water flow. The potential gradient for an ocean current or stream on a global

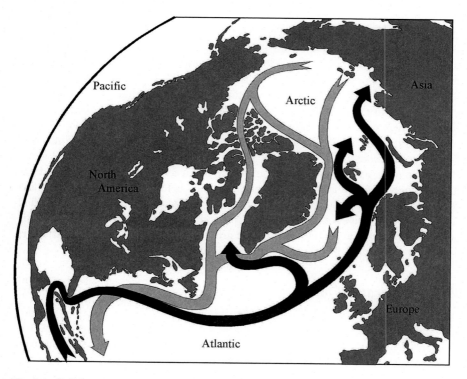

Fig. 9.1 Gulf Stream is a part of global ocean circulation – thermohaline circulation. It starts in the Mexican Gulf and flows to the north and to the northeast. It is warming up the parts of North America and of the Western Europe. Far on the north the waters cool and increase their salinity. Several branches of the stream flow therefore down to the ocean bottom and flow as bottom stream to the south

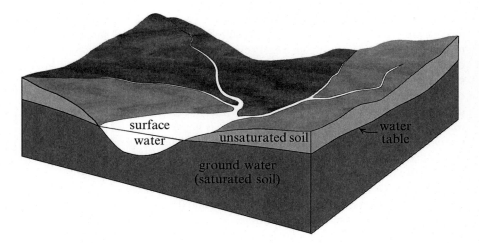

Fig. 9.2 Groundwater table or simply the water table separating saturated soil from the upper region of unsaturated soil. This boundary identifies saturated and unsaturated flow regimes that simplifies approximate calculations of practical problems. Water flows across this boundary either up at the evaporation or contributing to the supplying of plant roots by water. Or water flows down when rain contributes to the rise of groundwater level

or megascale is essentially comparable with the slope of the water level in a river if its riverbed is approximately straight, its length small, its flow rate small, and its temperature together with its content of dissolved salts does not change along the observed part of the river. In other words, when we observe and estimate the water flow in a river, we consider only the gravitational potential (Fig. 9.2).

Similar laws are valid for flow of water in soil, i.e., in porous media on a substantially smaller scale that is commonly called the "pore scale." Two principal water flow regimes in soil are usually designated. When a soil is water saturated and has all of its pores filled with water as typically occurs below a groundwater table, the flow is described as saturated flow. When only a portion of the pores in a soil is filled with water, we speak about unsaturated soil and unsaturated flow. Groundwater is separated from the unsaturated soil by the groundwater table or simply water table.

9.2.1 Saturated Flow

The most frequent type of saturated flow is the flow that occurs in groundwater. With the main driving force being the slope of the groundwater level, we measure it using two wells separated by a horizontal distance across which we wish to calculate the rate of groundwater flow. The wells are either dug out, or more frequently drilled, and provided with a floater. Just after digging or drilling an observation well, the water level inside of the well indicated by the floater rises owing to the inflow of water from the water-saturated region below the water table. After awhile the water level in the well no longer changes and remains constant. At this time we can start to measure and evaluate the flow of the groundwater. The difference of the water levels in our two observation wells divided by their horizontal separation distance is the driving force of the flow, or more exactly expressed, it is the potential gradient. We assume that the groundwater level has a smooth, continuous slope. The groundwater flows in the direction of this slope.

When we are in the valley of a river and the slope of the groundwater level leads from the river into the sediments of the valley, we know that the river is supplying the surrounding sediments and soils with water. This process may be especially important during a long-lasting rainless period of time if the vegetation in the valley can be adequately supplied with water coming from the river by groundwater flow. The accessibility of the river water to plant roots depends upon the depth of the groundwater level. We will deal with this aspect in detail later on. An opposite situation exists whenever the groundwater level is inclined from the surrounding alluvial sediments toward the riverbed. Groundwater would then flow from the valley sediments to the river. Hence, with the river being supplied by groundwater, the valley sediments are being drained by the river. We could easily verify this situation from our two observation wells because the water in the well close to the river would be lower than the water in the well more distant from the river. A combination of both situations also happens sometimes when one portion of the valley is drained by the river and another portion of the region is supplied by groundwater streaming from the river.

The flow rate of the groundwater depends upon two factors. The first is the slope of the groundwater level and we wrote about it. In practice for a given soil, the greater slope of the water level, the higher is the flow rate. We know already that if the words water level are replaced by the word *potential* and if the slope of the water level is substituted by *potential gradient*, we are more exact. Later on, we shall show the advantage and real benefit of using the second term when we write about unsaturated flow. Now back to saturated flow. The second important factor is the property of soil to allow water to penetrate into it or be conducted through it – a characteristic denoted by the expression, by saturated hydraulic conductivity, or sometimes simply by conductivity. It has the dimension of velocity (frequently m/day or cm/h) and is equal to the flow rate when the hydraulic slope is unity.

Basic equations were formulated in the middle of the nineteenth century by the French engineer Henry Darcy (1803–1858). After graduation at the most famous engineering school of that time, School of Bridges and Roads (*L'Ecole des Ponts et Chaussées*), he spent the majority of his life in Dijon, the town in Cote d'Or in the Burgundy region. Today, the region is famous for its wine. Much earlier when decimated by the plague, Dijon was well known in all of Europe for its extremely bad water quality. During that time period, Darcy decided to try to construct a new, modern, sanitary, and efficient water supply available to all inhabitants of the town. He studied theory of hydraulics and performed many experiments to test his new theoretical developments on flow of water in pipes throughout the town distribution system. Among other new approaches, he discovered the law how water flows in sediments and in sand filters used for improvement of river water quality. He considered the groundwaters in springs located in alluvial sediments as the best source of high-quality waters for Dijon. Putting all his knowledge into the service of water supply for the town Dijon, he succeeded and the town opened one of the most advanced water supplies at that time with many public fountains in 1844, a couple of years after completion of the project. In 1856, shortly before his death, he wrote the book *Public Fountains of the City of Dijon* (*Les Fontaines Publiques de la Ville Dijon*) to guide other engineers in constructing similar edifices of public importance. It was well before water supply systems were common in European or US cities. The equation and the basic law about flow of water in porous materials are named in honor of his legacy.

As mentioned earlier, hydraulic conductivity has the dimension of velocity (length/time or L/T). Unfortunately, we find less exact terms are sometimes introduced in nonscientific literature, especially, e.g., permeability and filtration coefficient. However, the scientific word permeability has an exactly defined meaning with the dimension of L^2 used for description of "flow" of air and generally of gas through porous materials, like soils and various filters. And the scientific term filtration coefficient is reserved for catching soil or solid particles bigger than is the size of pores.

Soil hydraulic conductivity depends upon several soil properties. Soil porosity and pore-size distribution are those most frequently cited for their dominating influence and control of its value. The saturated hydraulic conductivity of sands and sandy soils is very high because water flows primarily through their big pores with minimal viscous resistance because the water comes in contact with only a small solid area. The conductivity values of sands and sandy soils are usually about 100 cm/day. Loamy soils have lower values, usually several tens of centimeters per day, since they contain a combination of big and small pores. And, if small pores dominate their porosity, the conductivity sinks to about a centimeter per day or even less. The value of conductivity depends strongly upon the degree of soil aggregation. Loamy soils with well-developed, stabile structure may have a conductivity up to 50–100 cm/day, but when the structure is totally deteriorated, the conductivity decreases by as much as 50 times. Clays manifest the lowest values of hydraulic conductivity. With their specific surface being very large (50–200 m²/g), water moves with great difficulty owing to the very high friction losses at the boundary between solid clay particle surfaces and water. Conductivity of clay soils may fall well below 1 cm/day with their specific values strongly dependent upon the composition of the clay minerals (Fig. 9.3).

If a clay soil composed mainly of smectites is dry and cut by vertical cracks due to shrinkage, its hydraulic conductivity would be high during the first 10–60 min of infiltration. Subsequently, as a result of wetting and swelling, its conductivity would be almost nil. On the other hand, when a clay soil remains moist and never dry, it is considered as an impervious material frequently used for the internal sealing core of inundation dams that protect the plains from floods during high discharge in the river. Clay is used in a similar way in dams of reservoirs. In the middle of the dam constructed of compacted loamy soil, there is a trapezoidal cross section of a compacted clay core. Without this type of sealing core, water from the reservoir soaks through the dam and escapes to the downhill side of the dam. This leaking water

Soil texture	Sandy soil structureless	Loamy soil aggregated	Loamy soil structureless
Dominant particles	Sand	Silt	Silt

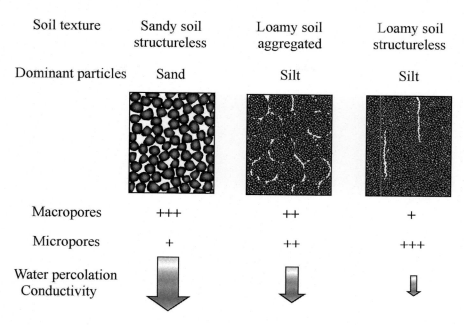

Macropores	+++	++	+
Micropores	+	++	+++
Water percolation Conductivity			

Fig. 9.3 Schematic role of soil texture and soil structure upon saturated conductivity and water percolation of the soil layer or of the whole soil profile. Sandy soils are either structureless or with a weak structure. The pores (indicated in *white*) between sand particles belong mainly to macropores and the hydraulic conductivity is very high. The well-aggregated loamy soil contains macropores between aggregates and micropores inside of the aggregates. Hydraulic conductivity is high to medium. Structureless loamy soils contain mainly or only micropores between soil particles. Macropores are either absent or they exist in minority if the soil is cracked when dry. Individual macropores may appear after decomposition of dead roots of plants. Hydraulic conductivity of the structureless loamy soil (except of cracks) is low

steadily loosens and removes particles of the loamy dam material and gradually decreases the stability of the dam until it finally ruptures.

When water flows through a system of soil layers where the conductivities differ, the overall total rate of flow depends upon the orientation of flow in relation to the arrangement of the layers; see Figs. 9.4 and 9.5.

The simplest procedure of measuring the saturated hydraulic conductivity is illustrated in Fig. 9.6. We measure the volume V of water percolating through the horizontal soil column of the length L during a period of time t. When we know the cross-sectional area A of the soil, we obtain the flow rate $q = V/At$ [(cm^3 water/cm^2 soil)/day] or simply (cm/day). The volume of soil taken in the field is usually not bigger than 1,000 cm^3, i.e., one liter. The greater the difference of water levels at the beginning and at the end of the soil column Δh, the greater is the flow rate q in the same soil. It looks logical and simple and this is the magic of Darcy's law whose equation, $q = -K_S \, \Delta h/L$, says that the flow rate depends upon the hydraulic gradient (or "slope") $\Delta h/L$ and upon the property of the soil to allow water flow – the saturated conductivity K_S. When the soil is sandy, water flows at very high rate through

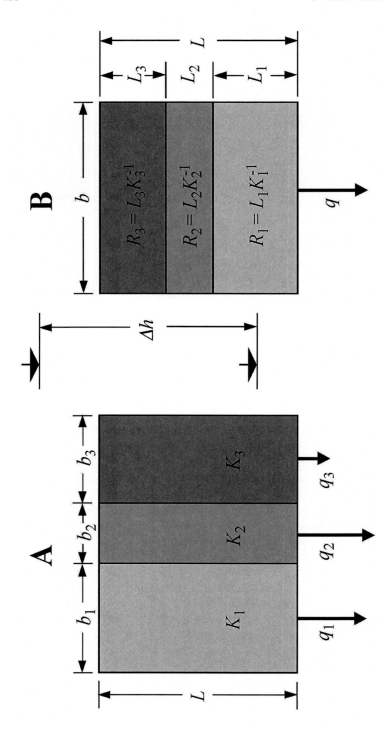

Fig. 9.4 The flux through three layers depends upon mutual orientation of layers and direction of flow. The total flux equals simply to the sum of fluxes through individual layers, if the layers are arranged in the same direction as is the flow, i.e., vertically in our system (**a**). If the layers are horizontal and the flow is vertical in system (**b**), then the total flux depends upon the sum of resistances R of individual layers and the value of the flux is much lower than in system (**a**)

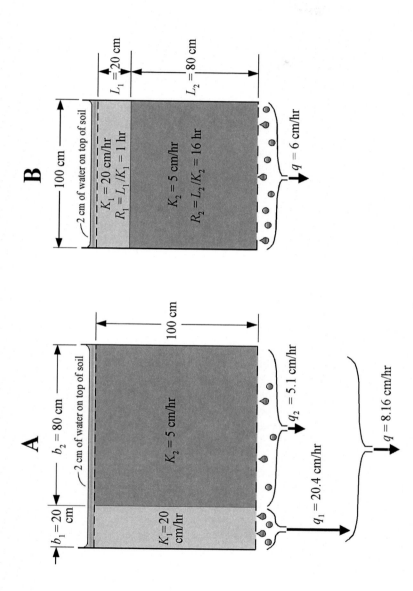

Fig. 9.5 A simple 2-layer example of the "theoretical drawing" of Fig. 9.4. The top water level is exactly drawn in the graph (2 cm above the land surface), the bottom water level at the outflow is here taken as identical with the bottom of the soil column. Hopefully, it is an easily visualized illustration of computing the outflow flux

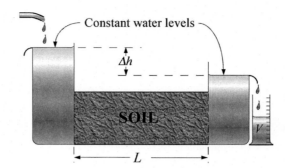

Fig. 9.6 The flow rate $q = V/At$ [(cm³ water/cm² soil)/day] where V is the volume of water caught after flow through soil column of cross section A and length L in time t, usually one day. The greater is the difference of water levels in this experiments the higher is q. The greater is the soil length L, the smaller is q. In order to get rid of those outside values, Darcy introduced the saturated hydraulic conductivity $K_S = q/(\Delta h/L)$

the apparatus in Fig. 9.6 because the value of K_S is high. For the same value of Δh when a loamy soil is in the apparatus, the flow rate is substantially lower because the value of K_S for a loamy soil is much smaller. And because the K_S of clay is extremely small, for the same value of Δh, the flow within clay could even look as if it were stopped or standing still.

A similar laboratory equipment for measuring saturated hydraulic conductivity K_S is in Fig. 9.7 with the total potential head $H = h + z$ where h is the pressure head and z the gravitational head measured from the reference level $z = 0$. Maintaining a constant level of water ponded on the soil surface causes water to flow steadily downward through the soil a constant rate q. With a measurement of this rate, the value of K_S can be determined from observations of total head H made at the top and bottom of the soil column or by making observations of pressure head h using piezometers placed at two vertical locations inside the column as illustrated on the left and right sides of Fig. 9.7, respectively.

Estimates of conductivity from field measurements provide more realistic information since their computed values relate to a natural soil area many times greater than that of small, disturbed samples measured in the laboratory. A frequent, reliable field procedure is to measure the decrease of groundwater level in the vicinity of a well being pumped and steadily removing water from saturated soil below the water table. Approximate analytical solutions or numerical simulations are used to obtain the saturated conductivity representing the groundwater flow in the region between a pumping well and nearby observation wells.

Equations and computation of saturated flow for solutions of practical tasks are relatively simple. They have been used for decades in computing and managing water from a system of wells for water supply and waterworks. Or, in agronomy, the drainage of waterlogged soils and the management of water tables to optimal depths are another example of the application of saturated flow equations for the benefit of practical life.

Fig. 9.7 A simple laboratory
equipment for measuring
saturated hydraulic
conductivity K_S

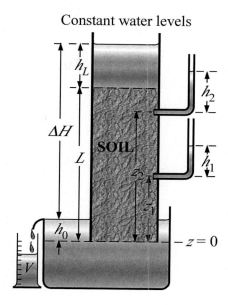

9.2.2   Unsaturated Flow

Although equations for saturated flow are relatively simple and straightforward, computing water movement in unsaturated soils is a complicated task even for a seemingly very simple situation. One of the several difficulties encountered is the fact that the hydraulic conductivity of an unsaturated soil does not have a unique value. Its value sinks significantly whenever the soil water content decreases by only a fraction of a percentage point. When the water content of an initially water-saturated soil starts to decrease, the biggest pores are the first to empty and cause an abrupt reduction of the hydraulic conductivity. We have already learned that big water-filled pores cause high values of saturated conductivity. However, when the soil water content drops below saturation, the majority of those big pores no longer conduct any water. And the lower the soil water content, the smaller is the size and total volume of water-filled pores that conduct water through the unsaturated soil. It is a common occurrence that the conductivity decreases by tenfold or even more when the soil water content decreases from saturation by only a very few percentage points. In other words, we would measure a value of K in a very wet, almost water-saturated soil to be only 10 cm/day compared with a value of 100 cm/day when it is fully saturated.

Complications between unsaturated conductivity and its relationship to soil texture never completely disappear as implicitly implied in Chap. 7 where the graph in Fig. 7.6 of the shape of an idealized log-normal soil pore-size distribution was presented while judiciously ignoring soil structure. Here in this and the next paragraph we continue to ignore complications of soil structure. Figure 9.8 illustrates relationships between the water-filled cylindrical pores of different sizes of radii r_i, water

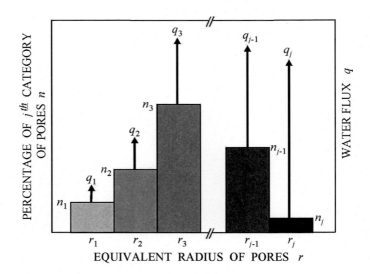

Fig. 9.8 Pore-size distribution in an ideal homogeneous soil with pore radius $r_1 < r_2 < r_3 \ldots$ (see the *rectangles*) and the fluxes in individual pore-size categories (see the *arrows*). The most frequent are pores here denoted by r_3, but the maximum flux qj is in the group of pores of maximum size, radius rj, in spite of their relatively low frequency

fluxes qi within each pore-size category, and percentages of each size category ni. The smallest and largest radii are r_1 and rj, respectively. Assuming the water flux qi for each category is a function of ri^4 and ni, it is obvious that a nonsymmetric water flux distribution exists inasmuch as the water flux at rj manifests a maximum value compared with all of those at smaller radii. As in many structureless soils, the pore-size distribution is approximately bell shaped with a peak of n_3 at radius r_3.

In the vicinity of water saturation, values of hydraulic conductivity of sandy soils are generally much higher than those of finer-textured soils. In contrast, as their soil water contents progressively decrease, values of hydraulic conductivity of finer soils are greater than those of sands. Consequently, a curve of hydraulic conductivity versus soil water content of a sand is not parallel to that of a loam. Indeed, as shown in Fig. 9.9a, their curves cross each other at a common value of soil water content θc. Similarly, as shown in Fig. 9.9b, their curves of hydraulic conductivity versus pressure head cross at a common value of pressure head hc.

Let us now consider the majority of structured soils. We find that curves of their pore-size distribution do not have just one peak but usually at least two peaks as demonstrated by Fig. 7.7 in Sect. 7.3. For such soils having a bimodal porosity, one portion of coarse pores occurring between aggregates (interpedal pores) has a peak for a radius somewhere between 20 and 60 µm and another group of fine pores occurring inside of aggregates (intrapedal pores) has a peak for a radius between 0.5 and 5 µm. When the soil is close to saturation, the flow of water is dominantly influenced by most of the coarse interpedal pores remaining filled with water and sustaining a high value of the unsaturated conductivity. When the soil is at lower water

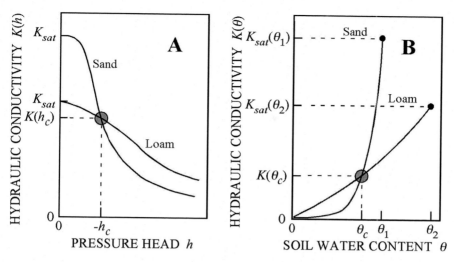

Fig. 9.9 The unsaturated conductivity of sandy soil is steeply falling down with small decrease of soil water content, compared to loam

contents, flow is maintained primarily by small pores (micropores) and the unsaturated conductivity sinks by an order of magnitude; see Fig. 9.10.

When a dissolved substance is transported with the flowing water and the soil water content is high, the solute is carried deep into the profile by fast-flowing water between the aggregates, while it is only negligibly transported into aggregates, since the conductivity within aggregates is many times, even a hundred times, smaller than that between the aggregates. This kind of flow in coarse pores is called "preferential flow" or "preferential transport," since it is preferentially restricted to coarse pores conducting water and dissolved matter at high transport rates through soils at or close to water saturation.

From our experience observing the behavior of soil water in natural outdoor environments, we are inclined to say that water flows in unsaturated soils from a place of high water content to one of low water content and additionally comment that the driving force of the process is the difference in soil water content. We often further state that the water content gradient determines the direction of flow and that its magnitude controls the rate of flow. However, the above statements are valid only for horizontal flow in the same type of soil with the same hydraulic characteristics, like saturated hydraulic conductivity, soil water retention curve, pore-size distribution, and compaction, and, moreover, with the same concentration of dissolved chemicals in soil pores. When these highly unlikely circumstances occur, the basic equations for this exceptional situation are similar to diffusion equations. For all other situations, we must adopt the concept of soil water potential as explained in Chap. 8.

When we observe the vertical flow of water in a soil that is homogeneous throughout its profile, at a specific time the soil water content together with its soil water potential may not vary within the upper 10, 20, or even more centimeters of topsoil –

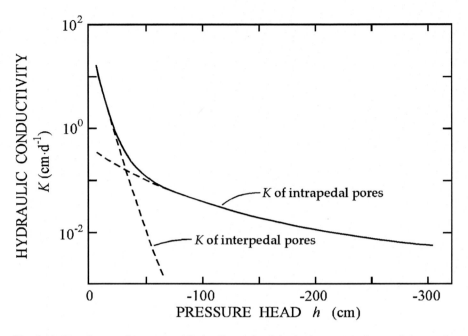

Fig. 9.10 The change of unsaturated hydraulic conductivity in the system of pores between the aggregates shows a strong decrease with a slight decrease of soil water content below the full saturation. In the system of fine pores inside of the aggregates, the decrease of unsaturated conductivity is gentle with the decrease of soil water content or the pressure head

a clear indication that the gradients of both soil water content and pressure head are zero. If no additional water flows through the surface of the topsoil (e.g., no rain and no evaporation since the surface is covered), we shall find after several hours or much more distinctly after several days that the water content at the 10-cm depth is smaller than that at depths of 30 or 40 cm. Indeed, water did flow down. From sequential measurements of soil water content and pressure head, we notice what Edgar Buckingham learned from his own experiments more than a century ago – gradients of gravitational potential acting together with those of soil water pressure potential provide the force responsible for vertical water movement through unsaturated soils, and values of hydraulic conductivity of unsaturated soils depend upon soil water content and as well as additional soil parameters and processes. Buckingham (1907) was the first person to significantly modify and adapt the original 1856 Darcy equation to quantitatively describe water movement through unsaturated soils. His seminal equation, nowadays frequently called the Darcy-Buckingham equation, constitutes a unique application of Fick's law of diffusion to compute the rate of water movement within an unsaturated soil. Professionally active for 50 years, his career was filled with an excellent mixture of diverse, unique achievements. Although highly appreciated by theoretical physicists, soil scientists trying to understand soil water behavior and its availability to agricultural plants essentially ignored all of his ideas. With their knowledge of mathematics and physics being inadequate to build upon his suggestions, they struggled for several decades just making empirical correlations between soil texture and field capacity or permanent wilting point. After working only 4 years in the US Department of Agriculture, Buckingham spent the remainder of his career in the US Bureau of Standards publishing many novel ideas including his well-known π theorem of dimensional analysis applicable to many scientific disciplines including those of soil science.

We described capillary pressure in Sect. 8.2. and soil water potential in Sect. 8.4. Since the surface pressure of a plane water level is higher than the surface pressure of the curved water level in the capillary (at complete wetting), the capillary potential is negative with its magnitude designated in units of kilopascals (kPa). One kPa represents the 10-cm height of a water meniscus in a vertical capillary tube. Next, we demonstrate simple examples showing that water flows from places where it has a higher potential to places where it has a lower potential.

We measured the soil water potential at the same depth of 60 cm in two experimental pits. The potential in pit No. 1 was −15 kPa and that in pit No. 2 was −30 kPa. Inasmuch as −15 kPa is greater than −30 kPa, before digging the pits, water was flowing horizontally from pit No. 1 to pit No. 2.

However, dealing with vertical flow, we have to add the gravitational potential to the soil water potential in order to obtain the total potential because the driving force is the total potential gradient. For example, at a depth of 30 cm below soil surface, our measurements reveal a soil water content of 35 % at a soil water potential of −20 kPa. At the depth 50 cm, our measurements revealed the same values of 35 % and −20 kPa. However, it does not mean that there is no water flow, since we have yet to add appropriate values of the gravitational potential. Taking the soil surface as the zero reference level of the gravitational potential, the total potential at the 20-cm depth is −23 kPa and that at the 50-cm depth is −25 kPa. And remembering that water flows from higher to lower values of total potential, we learn that at that instant water is flowing downward. Moreover, we obtain the rate of water flow by multiplying the gradient of the total potential with the value of the unsaturated conductivity. However, the soil water potential at the 50-cm depth increases with time until the gradient of the total potential is equal to zero and flow stops. At that time, the soil water content at 50-cm depth being higher than that at 30-cm depth, we also verify that soil water content differences are not sufficient to elucidate soil water flow.

With the soil water content changing due to the unsaturated water flow, the value of the unsaturated hydraulic conductivity also changes. Below we shall describe the footsteps of only a handful of physicists during the past century to heed the stimulating experimental and theoretical suggestions of Buckingham to consider the flow of water in unsaturated soil as a simple diffusion process. Because several aspects of diffusion differ from those of classical hydraulics, we continue with a couple of simple examples of diffusion that happen in everyday life.

While grilling a chicken in your kitchen with its door and windows closed, you suddenly become aware that you have overcooked it because your eyes tell you that it is a dry chunk of coal-like meat and your nose reveals its stinky odor everywhere throughout the enclosed kitchen. After you open the kitchen door, the bad odor steadily spreads into the corridor and other rooms where connecting doors are open, but all windows remain closed. The air in your apartment is stagnant, i.e., there is no draft, convection, or movement of air. Yet, in spite of the stagnant air, molecules of stink readily extend the undesirable stench throughout the apartment. The flux of those molecules is called diffusion – you observe the same process after you slowly place a spoon with salt into a glass of water. After a time all of the water will be salty without having mixed or stirred the salt and water together with the spoon. The physical explanation is based upon the natural-occurring thermal, never-ending vibrations of all molecules that cause them to diffuse and redistribute from locations of higher to lower or nil concentrations. In the above examples, the high concentration of stinking odiferous molecules from the overcooked chicken originating in the kitchen diffuses throughout the apartment, and the salt contained in the spoon initially dissolves and diffuses within the water to eventually make all of the water in the entire glass salty and have the same concentration.

Diffusion was first described by the equation known today as Fick's second law. Because its diffusion coefficient was often not a constant, the equation was modified and became the Fokker Planck equation. Since a nonconstant diffusion coefficient was uniquely comparable to our nonconstant unsaturated hydraulic conductivity function, the utility of a diffusion-type format to describe unsaturated soil water flow was an emerging possibility at the beginning of the twentieth century. Early formulations were initiated by physicists at the American school of Utah State University. It was not just by chance that the first steps were done in the arid, somewhat hostile Utah environment where most farming is realized with irrigation often using poor-quality slightly saline water. The economic use of sparse water as well as preventing harmful salt accumulation in the topsoil required experimental and theoretical research to derive an adequate theory and to prepare optimal instructions for farmers. Lorenzo ("Ren") Adolph Richards, a student at USU, extended what he learned from his physics professors with his own research to complete and publish his Ph.D. dissertation, "Capillary conduction of liquids through porous mediums," in 1931. Ren completed at that time the Buckingham's revolutionary act of shifting soil physics from empirical observation toward the exact language of Newtonian physics.

Even with the availability of Richards' equation, the 1930s through most of the 1950s remained uneventful regarding the application of theoretical equations to describe unsaturated water movement. This period of more than 20 years occurred partly due to the fact that Richards did not attract a group of students eager to extend and actually to try to find solutions of their teacher's equations. Additionally, the equation resembling the various diffusion equations (and even in a more complicated form) was not sufficiently applicable to observations of unsaturated water flow in field soils. The situation was also difficult due to the impossibility to measure soil water content without disturbing the soil or an ideal homogeneous soil column in the laboratory. It was so until the application of neutron probe method. Starting from the 1950s in the twentieth century, the mathematics of diffusion was extended and immediately the procedures leading to analytical and semi-analytical solutions were applied in studies on practical tasks on water flow in soils not fully saturated by water, e.g., by Wilford R. Gardner and his doctorate students. Don Kirkham concentrated another group around him, and he was the first to develop the solution of Richards' equation to define boundary conditions by the infinite series. We have already mentioned his name earlier in Chap. 8. Anyway, more about practical applications will follow in the next chapter.

Conditions in field soils at any location within the global landscape are complicated owing to the fact that the soil temperature oscillates differently to various depths during every day of the year within soil profiles having uneven soil water contents as a result of regional and seasonal precipitation. The soil hydraulic characteristics like contact angle and capillarity are dependent upon the temperature. A similar type of influence exists in unsaturated hydraulic conductivity. Hence, we must consider the impact of variable temperatures creating combined flows of both heat and water within soil profiles. Similarly in warm arid climate, the soluble inorganic salts accumulate in a salty soil horizon and they are transported and this transport may influence the flow of water. Thus, we are in a similar situation as we described for the Gulf Stream at the start of this chapter. The difference is in the scale, while for Gulf Stream it was in hundreds of kilometers, and in soils the distances are in tens of centimeters. But the main principle still remains.

Chapter 10
Soil Regulates Circulation of Water on the Planet Earth

The circulation of water on Earth is composed of several basic processes. Water vapor existing in the air condenses into small waterdrops that fall to the Earth's surface as rain or snow. The water either soaks into the ground or it flows on the surface into rivers – the majority of which empties into the ocean. Moreover, portions of infiltrated water flow to the rivers as groundwater. Water evaporates from oceans, from rivers on its run to the ocean, and directly from the soil surface or indirectly from the plants as transpiration. When water circulates back into the atmosphere as water vapor, it is ready to take another run within the above processes, collectively known as the hydrologic cycle. If there were no soil, infiltration would not exist. A net of dry riverbeds, called wadi in a desert, would quickly conduct rainwater. During the great majority of time, landscapes would be without water, similar to today's desert regions. Without any water on land, there would be no chance for plants to grow and animal life would not exist. Indeed, life on land would be restricted to some microbes resistant to dryness existing for the majority of times when rainfall does not occur. With rain seldom appearing, large territories without rain could exist for several years. Frequent rains happening several or more times in the month would occur only in narrow strips of land at the shore. The remaining regions of continents without soils that have no porosity to catch and store water in its pores during rainless periods would always remain dry and could not contribute to evaporation.

By retarding some of the rain falling on land, soil maintains a water supply that is absolutely essential for life on the land. Soil influences the rate of surface runoff and is a decisive factor together with the weather about the occurrence of floods. The deterioration of soil properties combined with variable weather conditions is responsible for occasional dry seasons that last long enough to produce extreme food shortages and famine in agricultural regions. We have no tools for making vital changes in the weather as, for example, how to bring rains during long-lasting dry periods or within specific seasons or years suffering excessive dryness. But we can change the soil properties either in a favorable or unfavorable direction for agriculture and for society in general. Being unaware of the final consequences of our

© Springer Science+Business Media Dordrecht 2015
M. Kutílek, D.R. Nielsen, *Soil: The Skin of the Planet Earth*,
DOI 10.1007/978-94-017-9789-4_10

activities, changes may occur even now against our desired goal. It happens when we are confidently persuaded that we are supporting sustainability without any proof that our actions improve the function of soil in the hydrologic cycle on local and global scales. Indeed, our actions may act in the opposite direction to worsen the cycle and conditions for life and society. It should be apparent that knowledge about soil hydrologic functions is unmatched and more vital than that about other ecosystem functions.

The capacity of a soil to catch rainwater is the same as that of all natural lakes as well as artificial lakes constructed by man for societal purposes. The amount of water kept in soils equals one-third of the total amount of those residing in all lakes. Comparing this volume to waters in all rivers, we find that the amount of water in soils is ten times more than all of the water in all rivers. If we add the volume of groundwater to the volume of soil water, we obtain a volume that is 100 times larger than the volume of all fresh surface waters in rivers and lakes. And when we study the circulation of water on Earth and its atmosphere, we learn that the role of soil is not negligible. Indeed, 74 % of global precipitation falls onto water and 26 % falls on land surfaces, while evaporation from water surfaces makes 81 % and that including transpiration from land surfaces is 19 %. The difference of 7 % between distributions of precipitation and evaporation from land surfaces is balanced by water running off soil surfaces.

Next, we are going to discuss details regarding the main hydrologic processes in soils – infiltration, evaporation, and transpiration – and also explore what happens within a soil between two infiltration events.

10.1   Infiltration of Water into Soil

Rainwater as well as water from melting snow or ice is essentially absorbed by the underlying soil at the Earth's surface. Accordingly, we speak about infiltration of water into soil. Water also infiltrates into soil from flooded areas or slight depressions where it accumulated by the surface runoff during the rain. Hydraulically, we have to differentiate two types of infiltration according to the source of water:

(a) Infiltration from a suddenly flooded soil surface being permanently inundated with a depth of water greater than that of the soil surface roughness. In a practical way, the free water level is more than 1 cm above the natural soil surface. It is theoretically defined as infiltration with a Dirichlet boundary condition (DBC). It means from time zero at the beginning of infiltration that a constant water content is kept on the soil surface. Here, we mean the surface soil water content remains water saturated. Furrow or check basin irrigation is solved using this type of infiltration boundary condition. The same condition is used to calculate infiltration into flooded plains along rivers.

(b) Infiltration from rain or sprinkler irrigation knowing its prevailing intensity or rate of water flow into the topsoil. Solutions of infiltration equations using this condition – known as the Neumann boundary condition (NBC) – provide descriptions of unsaturated flow within soil profiles.

In both instances we are dealing with two terms. One is the infiltration rate q_0 that is equivalent to a velocity, directly related to rain intensity and frequently measured as mm/min. Cumulative infiltration I is the second term. It is the sum of infiltrated water from the start of infiltration up to a specific time, related to the "depth" of the rain and measured in units of millimeter or centimeter. Let us mention that one millimeter of rainfall is the equivalent of 1 l of water per square meter.

10.1.1   Infiltration from the Suddenly Flooded Soil

Infiltration of water into a soil suddenly flooded with water is characterized by an initially very high infiltration rate q_0 many times higher than the saturated conductivity K_S, quickly decreases, and monotonically approaches K_S; see Fig. 10.1. From Chap. 9, we know that two forces are involved – one varies with time, while the other is constant. The force of the pressure potential gradient is extremely high when the soil is initially dry and wetting by infiltration but diminishes to nearly trivial values compared with the force of gravity as the topsoil wets to nearly the same soil water content within the topsoil. We can compare it to the thirst of a man who did not get a drop of water during a long, hot day. He very quickly drinks the first gulp, but gradually as he is quenching his extreme thirst, he drinks slowly and

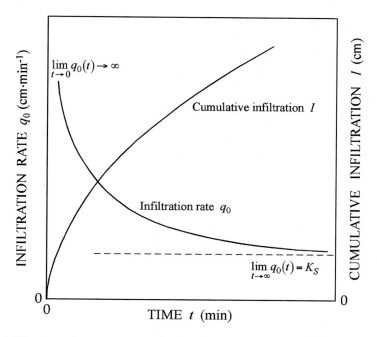

Fig. 10.1 Time dependence of the cumulative infiltration $I(t)$ and of the infiltration rate $q_0(t)$ when the soil surface is suddenly flooded, i.e., at flood infiltration, when the soil is not fully saturated by water before infiltration and when the soil is relatively homogeneous in vertical direction. After a very long time, the infiltration rate q_0 is roughly equal to the saturated hydraulic conductivity K_S

steadily. So it is with infiltration rate – its extremely rapid value quickly decreases with time of infiltration to eventually approach the value of the saturated conductivity. Similarly, the increase of the cumulative infiltration I is high at the start, but after several hours it slows down and reaches a constant value of increase.

During infiltration into homogeneous soils, those without layers or horizons, if we measure the distribution of soil water content θ within the profile at several different times and draw a smooth curve through the measured data, we discover that the water penetrates into the soil like a piston. The bottom of this "piston" is the wetting front of infiltrating water. From the soil surface down to just above the wetting front, saturated values of soil water content prevail with all pores being filled with water. Only within a very narrow range of depths immediately above the front does the water content sink below saturation. Although such profiles are commonly observed in sandy soils, infiltration profiles within loamy and clay soils do not manifest such a vivid resemblance to a piston; see Fig. 10.2. Profiles in this figure illustrate soils with simple pore-size distributions when the role of soil structure is of minor consequences.

Soil structure alters the size and distribution of pores in the majority of loamy and clayey soils. Their soil water profiles during infiltration are less regular than those illustrated in Fig. 10.2. Simply saying, deviations occur due to the more complex, irregular nature of their pores; see Sect. 7.3. Water penetrates these structured soils mainly and preferentially through sequences of their big pores that we previously described as preferential flow. The flow rate in these big pores is even more than ten times faster than that in the majority of fine pores existing mainly inside aggregates. This great difference of infiltration rate cannot disappear when the wetting front penetrates to a deeper soil horizon (below A horizon) even if there the conductivity of the "structureless" soil does not differ from the "structured" soil. The terms structured and structureless soil are used as a concise description of the topsoil (A horizon). Let us recall our description of fluxes in layered soils in Chap. 9, Figs. 9.3 and 9.4.

If a soil has a well-developed structure in the top horizon with stable aggregates keeping their shape even after being soaked by water, the infiltration rate slowly and steadily decreases with time. On the other hand, if the soil contains only quasi-stable aggregates that slake abruptly into slushy mass just after wetting in the top horizon, the infiltration rate sharply decreases. The differences between q_0 values for structured and structureless soils are about tenfold during initial stages of infiltration; see the Fig. 10.3. The slaking of crumbs is more drastically increased when a nonsaline water infiltrates a saline soil with an abundance of monovalent Na^+ cations.

10.1.2   Rain Infiltration

Initially, infiltration from rainfall depends upon the rain intensity (velocity) expressed usually in millimeters per minute. Let us first assume that the soil is homogeneous without any distinct layering. The infiltration rate is limited by the

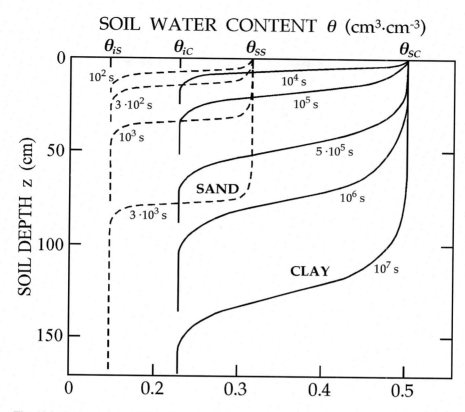

Fig. 10.2 Soil water content development $\theta(z)$ at certain time intervals of water infiltration into sandy soil (subscript S) and a clayey soil (subscript C) when the soil surface is suddenly flooded at time $t=0$. With initial and saturated soil water contents being denoted by θ_i and θ_s, respectively, the initial water content of the sandy soil is θ_{iS} and that of the clayey soil is θ_{iC}. Similarly, the water-saturated values are θ_{sS} and θ_{sC}. The numbers at the curves denote the time in seconds after the start of infiltration

"inflow" of water from the rain with the rain intensity distinctly lower than is the ability of soil to imbibe or soak up water after the start of infiltration, as we earlier demonstrated this theoretical ability for flood infiltration in Fig. 10.1.

During a rain of any intensity, various shapes of soil water content profiles develop sequentially in phases. Although they roughly resemble those of flood infiltration in Fig. 10.2, there is a very important and distinct difference that becomes evident when rain comes down at a constant intensity or rate. For an example, we consider a heavy rain falling at a constant intensity q_R that is several times higher than the value of the saturated conductivity. The soil rapidly becomes progressively more wet but never gets completely water saturated even at the top of its surface during the first phase of infiltration; see curves labeled t_1, t_2, and t_3 in Fig. 10.4. Just after the start of infiltration at time t_1, the soil water content of the surface is only

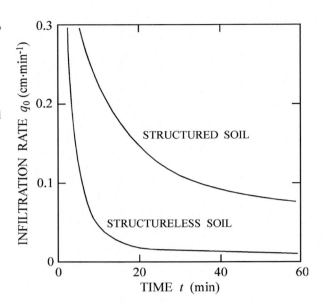

Fig. 10.3 Infiltration rates q_0 from flooded land into a well-aggregated loamy soil decrease more slowly than those into another loamy soil that lacks structural properties. The infiltration rate q_0 into the structured soil is up to ten times larger during the first 30 min. The aggregated soil maintains its substantially higher infiltration rate even after more than 1 h in our graph. The terms structured and structureless soil denote the properties of the top A horizon, only

somewhat higher than it was before the rain started. At times t_2 and t_3 as the rainfall continues, the soil water content at the soil surface continues to increase. In a little more time, it reaches θ_S – the saturated soil water content – that is at the top of the curve labeled t_4 in Fig. 10.4. Because this time is very significant and linked with other events during and after infiltration, it is equated to t_p and called the ponding time – the instant that free water starts to exist and form a pool on the soil surface or starts to flow across the topsoil.

We have compared soil water profiles during flood infiltration to a piston systematically penetrating into the soil profile with time from the very beginning of the process (Fig. 10.2). If we plot the soil water profiles during rain infiltration, we can use the comparison to a piston only with the provision that the piston gradually increases its horizontal diameter until the surface soil water content reaches saturation at the ponding time t_p; see Fig. 10.4. For times larger than t_p, the profiles during rainfall resemble those flooded profiles in Fig. 10.2 – both look like pistons of constant diameter penetrating deeper and deeper into the soil profile.

The term ponding time and its decisive role follow from Fig. 10.5, where the infiltration rate is plotted as dependent upon the time. Infiltration rate is constant in the first part since the source, the rain, has a constant intensity that is smaller than the ability of soil to swallow water from the surface. This situation lasts up to the ponding time. This is the moment when the ability of soil to soak water equals the rate of incoming water, i.e., the rain intensity. After this ponding time is reached, the ability of soil to consume all incoming water is smaller than is the rain intensity causing some of the rainwater to remain on the soil surface. Hence, the soil surface is ponded by water and the longer is the ponding time, the greater is the difference between rain intensity and the rate of actual infiltration. During this ponding time, the height of water remaining on the soil surface continues to increase provided that

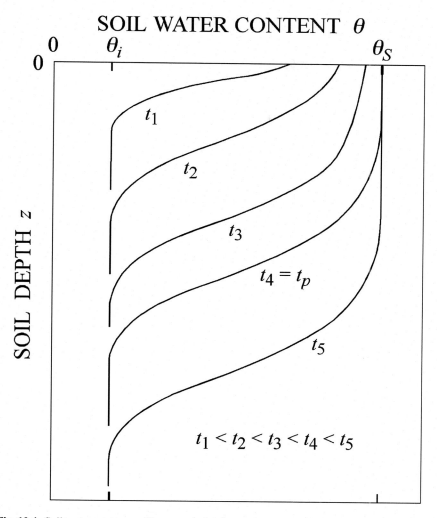

Fig. 10.4 Soil water content profiles measured at increasing times in a homogeneous soil initially at a soil water content θ_i during a rain of constant intensity. Time t_1 is just after the start of infiltration. The time t_4 equal to t_p designates the ponding time when the soil surface becomes water saturated θ_S. For greater times like t_5 in our figure, the curve shows the water-saturated segment of the profile continually elongating roughly in a similar way like in flood infiltration demonstrated in Fig. 10.2

the terrain is ideally flat. If there is a slope, the water flows down the surface with the runoff increasing with time. The water flowing on the surface starts to be concentrated in rills and finally it reaches a river or lake. If the rising water level reaches and exceeds a critical geographical level, a flood covers the valley of the river. Although the curves in Fig. 10.5 demonstrate that the ponding time comes earlier when the rain intensity is increased, we must remember that the hydraulic

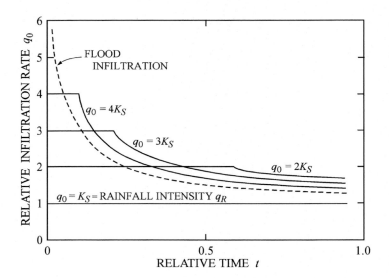

Fig. 10.5 Time dependence of the infiltration rate of rainwater when the rain intensity q_R is constant and greater than the value of the saturated conductivity K_S. The relative infiltration rate in the graph $q_0 = q_R/K_S$. At the beginning of the rainfall event, the relative infiltration rate q_0 is constant and equal to the rain intensity q_R until the soil starts to soak water more slowly than q_R. At that time, the ponding time t_p, any water that does not infiltrate starts to remain on the soil surface. If there is a slope, runoff starts. For rains that continue to last longer than the ponding time, the infiltration rate decreases and the shape of the time-dependent curve resembles that for a soil flooded with water (see Fig. 10.1) shown here as a *broken line*

conductivity property of the soil surface is always jointly responsible for the value of ponding time. Having already shown in Fig. 10.3 great differences in flood infiltration rate between structured and structureless soils, ponding time is influenced similarly by the quality of soil structure, especially at high rain intensity. The ponding time for a structured soil may be up to 10–20 times higher than that for the same loamy soil without any aggregate stability.

The real situation of rainfall and the consequent infiltration are usually more complicated with the determination of the ponding time and amount of infiltrated rainwater ("effective rain" in the language of hydrologists) being a matter of numerical modeling. But the basic principles discussed above are valid. The most important are two characteristics: (1) water infiltrates into the soil in the form of a slightly deformed piston, and (2) a specific time when the entire amount of rain no longer infiltrates into the soil and the non-infiltrating water starts to pond on the soil surface – a characteristic called ponding time. Its value is indicative of the beginning of surface runoff and the potential onset of floods.

On several occasions while introducing the principal characteristics of infiltration in relation to a water regime on a regional scale, we have mentioned the uniqueness of soil structure. Because the role of soil structure and aggregate stability cannot be replaced by any technological procedure, it is the moral obligation of society to sustain and even improve well-developed soil structure for the benefit of agricultural development, river hydrology, and generally for healthy ecologic parameters within the entire regions. Infiltration declines as soil structure deteriorates as a result of soil humus decreasing together with diminishing contents of humic acids, humins, and especially glomalin; see our earlier Sect. 6.2. Soil aggregation is easily deteriorated by improper soil tillage and by repetitively planting the same crops in monocultures. During the 1900s in the central plains of the USA, continuous planting of wheat not just for several years but for decades resulted in the wholesale destruction of soil structure followed by mediocre, unproductive water management. With a few years of low precipitation, inadequate rainfall infiltration resulted in an extreme drying of soils. Fine soil particles detached from the structureless soil blown by winds and dust storms were transported as heavy clouds depositing the hot silt and clay in leeward sites like huge snowdrifts, but instead of snow, hot fine soil particles were burning the leaves of plants. When the long-expected rain finally arrived, the infiltration of rainwater was practically absent because the soil surface was so dense and muddy. This impenetrability at the soil surface magnified the lack of water in the soil profile and even in a short time after the rain, the vegetation continued to permanently wilt and die from thirst. The excess of non-infiltrated water resulted in water ponding, surface runoff, and soil erosion on even a slight slope. The long-lasting monoculture planting resulted in the exhaustion of several important nutrients and the catastrophe was thus complete.

10.2   What Happens After Infiltration (About Soil Water Redistribution)

After infiltration ceases, we detect a gradual decrease of the soil water content in the top layer wetted by rain or flood even if the soil surface is protected by a cover which does not allow evaporation. The decrease of the soil water content in the originally piston-like wetted topsoil is caused by the downward flow of soil water. As water drains from and through the wet topsoil and is conducted to soil below the infiltration front by the gradient of the water potential, a new quasi or secondary wetting front is formed. The width of the piston-like wetted topsoil starts to be slimmer and slimmer, and the new secondary wetting front proceeds deeper and deeper; see Fig. 10.6. This transport of water inside the soil profile after infiltration is called soil water redistribution. It contributes to an optimal source of water within the entire root zone of plants since their roots may reach greater depths than the position of any infiltration front. Such soil water redistribution has other benefits – it opens water-filled coarse pores for air to penetrate into the newly created nearly water-saturated topsoil to reduce the danger of putrefaction of plant roots. It also opens and prepares the soil profile to benefit from the next incoming rain.

Soil water redistribution after each infiltration event eventually slows down in a complicated hysteretic manner until a constant value of the soil water content exists within the wetted portion of the soil profile. The water held in the moist profile attracted early studies on soil physics and was called "field capacity." During the

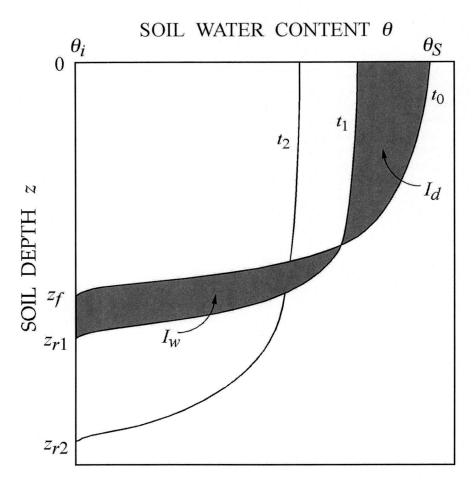

Fig. 10.6 Soil water content profiles due to redistribution at the end of infiltration. This time is denoted by time zero, t_0, when the depth of infiltration front is denoted by z_f. Then there is no water inflow at the soil surface, but the water content is decreased since the top wet part is drained and the water from this drained part flows downward to the originally dry subsoil. The volume of drained water from the infiltration I_d equals the volume of water wetting the soil below the infiltration front I_w, i.e., the volume of water in redistribution. All is valid provided that evaporation from the soil does not exist. Here again $t_0 < t_1 < t_2$

time of those first studies at the end of nineteenth century, the term "capillary capacity" was also used; see also Sect. 9.1. There were relatively simple methods on how to determine the value of "field" or "capillary" capacity in the laboratory on undisturbed core samples having a volume of about 100–200 cm³. Soil samples, each kept in a rigid cylinder, were first completely saturated with water and subsequently drained either on a thick layer of dry sand or on multiple filtering paper. The amount of water retained after such a simple drainage procedure was considered a reliable measure of field or capillary capacity. Since the empirically found rough estimate of the water potential at field capacity was 1/3 bar, the method of its estimation was modified accordingly. Field capacity is even now estimated in a pressure plate apparatus (Fig. 8.10) as water content at the overpressure of 1/3 bar.

10.3   Soil Water Evaporation and Evapotranspiration

In order to simplify these natural processes where in the great majority of instances evaporation directly from soil is accompanied by evaporation directly from plants, named transpiration, we are going to describe both processes separately. It will be easier not only for our discussion, but from a practical point of view, we can show how far we could reduce this "loss" of soil water (evapotranspiration) just to evaporation as one of the basic processes in the hydrologic cycle.

10.3.1   Evaporation

There are two different situations of evaporation from soil. First we suppose that there is groundwater level relatively close to the soil surface and the evaporation from the soil is actually supplied from the groundwater level. Another term commonly used to refer to a groundwater level close to the soil surface is water table. Evaporation of water from the soil E is indicated as a rate, usually in millimeters per day analogous to meteorological data or to infiltration rate. Its value depends upon the depth of groundwater level and upon the meteorological conditions of temperature, wind speed, air humidity, and solar radiation. Under a specific set of meteorological conditions, it is approximately equated to the evaporation rate from free water if the groundwater level is close to the surface.

More precisely, we can speak about potential evaporation or evaporativity as a meteorological characteristic. The deeper is the groundwater level, the lower is the soil water content in the surface horizon as well as its unsaturated conductivity. Or we can also say, the deeper is the groundwater level, the higher is the resistance against water flux for similar soils that differ only by the depth of their groundwater levels. It follows, therefore, that the deeper the groundwater level, the lower is the actual evaporation from the soil at the same potential evaporation; see schematically Fig. 10.7. The main simplification in this first situation is the constant position of groundwater level which should not sink deeper due to long-time evaporation.

The second situation, entirely different from the first, occurs whenever the groundwater level is so deep that it does not influence evaporation. If the soil surface is not covered by a folio or a dense cover of plant debris, the evaporation of water from a soil surface is accompanied by soil water redistribution with the water content decreasing sharply within the thin topsoil layer of about 1 cm thickness. We can observe this change of water content with our naked eyes because the darker color of wet soil changes to a lighter color. After a period of infiltration when the soil is

Fig. 10.7 The actual
evaporation from a bare soil
with a groundwater level
(GWL) close to surface
depends upon the depth of
GWL and upon the potential
evaporation. The deeper is
GWL, the lower is actual
evaporation if we compare
values at a certain potential
evaporation. If the GWL
reaches to the soil surface,
then the actual evaporation
equals the potential
evaporation

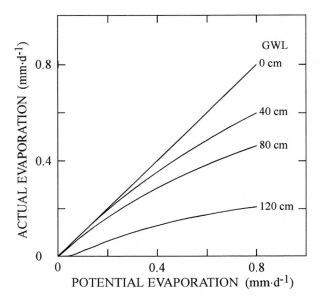

very wet (at or near saturation), the evaporation rate E is initially determined by
potential evaporation. However, the rate is then later reduced due to two facts: there
is a reduction of the amount of water freely accessible to evaporation since a major
portion of the horizontal cross sections of the profile is occupied by solid soil particles and aggregates. The second factor is the resistance opposing water vapor flux
as water molecules struggle through soil pores in their penetration upward to the
soil surface from positions below the soil surface. A simplified process is thus composed of two stages. The constant rate after the start of evaporation from saturated
soil is called the first stage of evaporation. It is equal to potential evaporation. It
ends at a time when a very thin top layer is gradually dried out and reaches a certain
low value of water content. After that time the second stage starts with a steadily
decreasing rate of evaporation; see Fig. 10.8. During this second stage, the depth of
the overlying dried layer increases.

The oversimplified situation illustrated in Fig. 10.8 is never fully observed in
natural conditions. Nevertheless, the first stage of evaporation is an approximate
reality just after the occurrence of long-lasting rains when the soil is saturated by
water to depths of about 30 cm and more. It is an approximate reality because
natural-occurring evaporation is never absolutely constant with time owing to the
fact that potential evaporation is never constant during day or night. The day/night
change of evaporation rate is high when the soil has a high water content (at field
capacity and more) in the top layer reaching from the soil surface to the depth about
20–30 cm. The day/night change of evaporation rate is low when the soil water
content is low (close to wilting point – see Sect. 10.3.2 and Fig. 10.11).

The evaporation rate is greatly influenced by the soil structure of the top horizon
especially in the second stage. If the topsoil has a stable well-developed structure,

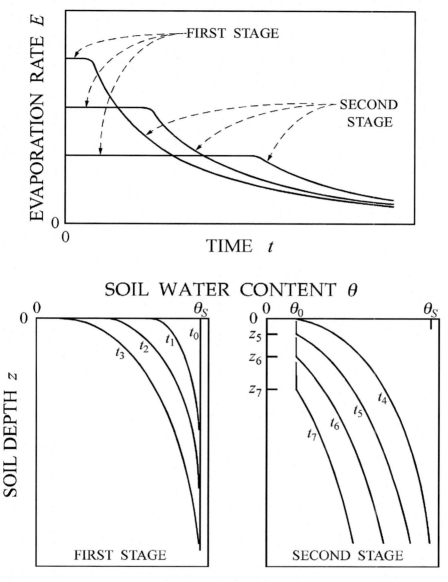

Fig. 10.8 An illustration of a rather artificial situation when meteorological conditions decisive for potential evaporation (evaporativity) remain absolutely constant during a long time for a soil initially water saturated (θ_S) throughout its profile. The evaporation rate from the soil during the first stage of evaporation is constant with its value depending only upon meteorological conditions (*top graph*). During this stage, although the soil water content at the soil surface readily decreases, it remains close to saturation at depths not very far from the soil surface (*left part* of *bottom graph*). The second stage of evaporation rate begins at time t_4 (*right part* of the *bottom graph*) with an abrupt decrease and followed by a gradually less decreasing rate illustrated in the *top graph*. During the second stage, the soil water content at the surface remains at a critical minimum value θ_0 that moves downward to form an increasing thickness of dry topsoil as evaporation continues (*right part* of the *bottom graph*)

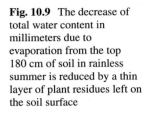

Fig. 10.9 The decrease of total water content in millimeters due to evaporation from the top 180 cm of soil in rainless summer is reduced by a thin layer of plant residues left on the soil surface

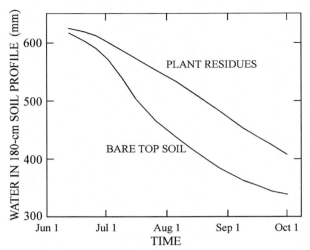

the coarse pores between the aggregates promptly lose water very early in the first stage of evaporation. With continuous paths for liquid water flow being broken, the final segment for water moving upward to the soil surface is realized by diffusion of water vapor. With diffusion causing a substantially smaller flux than that of liquid water, the evaporation rate decreases sharply.

A structureless soil surface is characterized by small clods that quickly disintegrate as they are pelted by a downpour of raindrops. With the surface changed into mud during each rainfall event, portions of it always proceed deeper into the soil profile. It is not at all unusual that a dry crust containing only small pores has been formed on a structureless soil before it rains again. The upshot of a continuity of small pores is their high conductivity compared with vapor diffusivity between aggregates in structured soil. Continuous liquid flow up to the surface in a structureless soil results in a higher flux than the flux in a structured soil, especially during the second stage of evaporation.

Soil tillage causes a short-term increase in porosity, but this positive effect is accompanied by a long-term decrease of aggregation that eventually ends in a structureless soil. This unfavorable situation has been solved by relatively frequent fallowing or by introducing clover and many other leguminous plants into crop rotations. Additionally, classical plowing is either restricted or replaced by a sort of disking using various types of chisel and moldboard plows. Such disk plowing keeps substantial quantities of crop residues on the soil surface that restrict the loss of water from the soil profile. Because a soil layer without a continuous water flux allows only vapor diffusion, evaporation is lowered. We see in Fig. 10.9 that the total content of water in a soil profile does not sink as sharply during the summer month of July as it does in the same soil without this protective layer of residues. In spite of the many recommended treatments and management techniques, the complete return of a structureless soil to its original structure with stable aggregates is a difficult task.

10.3.2 Transpiration

The long-lasting existence of bare soil is rare owing to the management of various types of landscapes. In agricultural regions of crop production, bare soils exist only after harvest, and even there, the plant residues are frequently left on the surface as a protection against excessive loss of soil water by evaporation as seen above in Fig. 10.9. Although we are not speaking about deserts and semideserts, in such regions we often observe irrigated crop production where bare soil may also exist. Water conducted by the irrigation system to the plants usually wets the soil with a portion of the applied water lost by evaporation from channels, furrows, and bare soil. Recent irrigation systems, particularly those called drip systems, are the most efficient and reduce water losses to a practical minimum. Water, conducted just to the plants by a system of tubing, is dripped directly to specific locations within the root zone of plants at selected times and rates of application.

We now return to geographical regions supplied by rainfall in a more or less sufficient way. Cultivated plants shadowing the soil surface of agricultural fields reduce evaporation from the bare soil surface. As the crop root system pumps water out of the soil, the thickness of soil layer "donating" water increases as plants grow and their roots more thoroughly penetrate the soil profile. Although nearly all of the water sucked up by roots flows through a plant to its leaves to participate in the formation of new tissues by photosynthesis in the presence of the sun's radiation, the vast majority of the water moving through the plant simply evaporates directly into the atmosphere from its leaf surfaces (Fig. 10.10). We easily compare this energized photosynthetic activity to an industrial process with evaporation serving as the cooling process within the factory. As in most simplifying analogies, the entire process is more complicated and requires potential gradients to enable the upward flow of water. This form of evaporation from plants is called transpiration.

The ratio of the mass of water extracted by a plant from the soil in order to produce dry organic matter of the plant is denoted as the *transpiration coefficient* (or *ratio*); it is simply expressed as grams of water to produce 1 g of plant organic matter. Plants need much more water than animals for production of 1 g of their bodies.

The transpiration coefficient depends on climatic and soil conditions and on the species of plant. Its value ranges from 300 to 900 for cultivated plants and from 150 to 700 for trees of the temperate zone. Water is the main constituent of plant tissues. Although the percentage of water in hydrophytes (aquatic plants growing only in water) is about 80–85 % and a little less in cultivated plants (70–80 %), the water content of individual cells or particular plant parts deviates greatly. The metabolically most active plant parts and young tissues contain about 90 % water in relation to dry organic matter, leaves of cultivated plants 70–90 %, wood of trees about 50–60 %, and seeds as little as 5–10 %. About 95 % or even more of the water extracted from soil is transpired and flows through the soil-root-stem-leave-atmosphere system. Almost all of it winds up in the atmosphere by escaping through an exorbitant number of microscopic windows in the leaves called stomata. A very

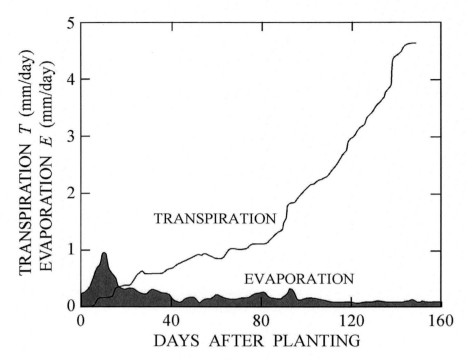

Fig. 10.10 Rate of transpiration and rate of evaporation directly from the soil surface measured in a rape field during the vegetation season. Except during the first 3–4 weeks after planting, transpiration exceeds evaporation from the soil surface because the rape canopy shades the soil surface

small portion of the water moving within plant stems also escapes into the atmosphere through a few stomata as well as other pores called lenticels that serve as windows in the external stem tissue. Because stomata can be opened, half closed, or completely closed like real windows, plants are able to regulate their transpiration.

The driving force of the water transport is the gradient of the water potential. Resistances at the soil-root and leaf-atmosphere interfaces and those inside the plant, soil, and atmosphere influence the actual fluxes; see Fig. 10.11. The water potential in plant tissues is defined analogously to that in soils. The total plant water potential is usually partitioned into two components: (1) turgor or turgor pressure identified with a pressure soil potential and (2) osmotic pressure.

An increase of turgor is accompanied by an increase of cell volume and a decrease of osmotic pressure to zero at full turgor. A decrease of turgor owing to dehydration causes the cells to shrink and at a certain threshold value we observe the wilting of leaves. Complete wilting occurs at osmotic pressures between 0.5 and 20 MPa. Roughly one century ago, the first generation of soil physicists started to use the average value and called this characteristic the wilting point, as we mentioned already in Sect. 9.1. We describe its main mechanism here with an understanding of plant physiology.

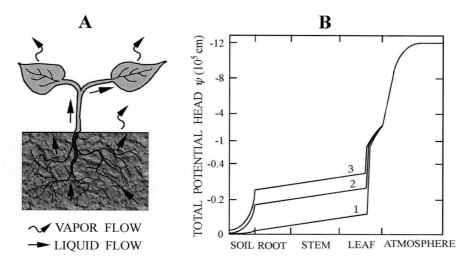

Fig. 10.11 (a) Schematic demonstration of liquid water flow through soil, roots, and plant leaves from which transpiration exists as water vapor flow. (b) *Curve 1* is the distribution of potential values within the soil-root-stem-leaf-atmosphere continuum when the soil water content is close to field capacity. *Curve 2* is when the soil water content is well below field capacity. *Curve 3* is for a relatively dry soil when the water content around roots is at wilting point

The total potential at permanent wilting can manifest a broad range of values that depend upon the species and stage of development of the plant and the local environmental conditions. This extreme decrease of potential is accompanied by irreversible morphological changes inside of the cells that stop the growth of the plant.

Water enters plants primarily through root hairs and the most active roots – both young and old – depending upon their distribution within local soil water availability regions of the soil profile. The main "riverbed" of water transport in plants is denoted by the term xylem. After entering the roots, water moves through the xylem against a very small resistance, enters the stem, continues to flow, and easily enters the leaves – the dominant locality of transpiration.

A simplified cross section of a leaf is illustrated in Fig. 10.12. The surface of a leaf is constituted by a thin impermeable film (an epidermis covered by a waxy cuticle). Micro-openings, micro-holes, or micropores earlier identified as "windows" called more precisely stomata exist at an average density of about 50–500 holes in each epidermal area of 1 mm². To better appreciate these densities, we return to Fig. 10.10 to ask ourselves, how many stomata are on each rape leaf in that farmer's field? The answer: the average rape leaf contains as many as 11,000,000 stomata. Stomata occur on both surfaces of the leaf of cultivated plants, but they often occur more commonly on the lower surfaces of tree leaves. The shape, opening, and closure of each stoma are regulated by its two kidney-shaped guard cells. With the turgor in the leaf cells transferred to the guard cells, a decrease of turgor deforms the guard cells up to closing most all stomata when we observe the wilting of leaves. With transpiration reduced to nearly zero and no substantial quantities of

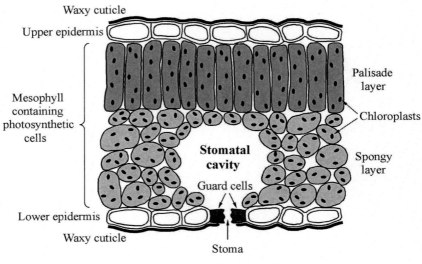

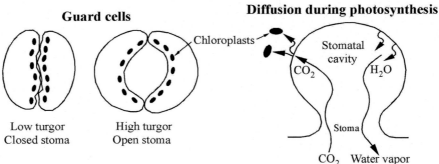

Fig. 10.12 Cross section of a leaf, closed and open guard cells, and exchange of CO_2 and H_2O by diffusion within a stomatal cavity during photosynthesis

CO_2 entering stomatal cavities, photosynthesis comes to a halt. An increase of turgor functions in an opposite way to open the stomata. The aperture of each stoma forms a pathway for diffusive transfer of water vapor. However, the function of the guard cells including stomatal opening and closing is not a simple process that reacts merely to cell turgor and relative conditions of plant water stress. Sensors below each stoma react to concentrations of CO_2 in the air, radiation, temperature, etc. For example, the presence of toxic gases in the air causes stomatal closure. Potassium ion concentration and proton transport across the membrane of guard cells play a dominant role regarding the dynamics of opening and closing of stomatal apertures.

The primary self-regulating mechanism of transpiration by a plant is realized through stomata. With a gradual closing of stomata, the resistance against water vapor flux increases and transpiration decreases. However, transpiration does not stop even if the stomata close completely. And if transpiration exceeds the influx of water into the roots, plant tissues lose both water and turgor. If the loss of turgor continues beyond a critical threshold value, wilting occurs. Wilting can occur even if the soil is fully moist when the evaporative demand of the atmosphere is large and the influx to the roots is lower than the transpiration loss. Often during a hot summer midday, plants having large leaf areas may temporarily wilt. Subsequently, in the evening hours when the extreme air conditions cease and the transpiration rate decreases, turgor in the leaf tissues returns. When extreme air conditions prevail or when the soil remains relatively dry for several days, the loss of turgor in cells accompanied by irreversible changes in the plant finally causes permanent wilting. Stomatal apertures also provide pathways into the plant for CO_2 needed for photosynthesis and the growth of plant tissues. Hence, a correlation between transpiration and crop yields exist and has been frequently documented. Some authors claim that the priority for receiving CO_2 necessary for photosynthesis indeed regulates stomatal aperture.

Transpiration rates of a certain type of plant and its variety depend widely upon soil water content and upon weather conditions, such as temperature, humidity, and sunlight intensity. Analogous to potential evaporation, the potential transpiration is defined as the loss of water from plant tissues to the atmosphere according to the evaporative demands of the atmosphere with the stomata fully opened. Additionally, it is understood that water movement from the soil to the plant roots does not limit the process. With atmospheric conditions including the energy source and radiation controlling the phase change of water, energy regulates the process for a given stomatal density. The term unstressed transpiration is also used for potential transpiration.

Let us now assume that the soil water content suddenly decreases. Concomitantly, values of soil water potential and unsaturated hydraulic conductivity both drastically decrease. In such a case, the evaporative flux is maintained by the increase of the gradient of the decreasing soil water potential. As a consequence the cell water potential continually decreases and is accompanied by a loss of cell turgor and at its critical value, the stomata start to close. This critical value also depends upon plant type, its variety, its susceptibility to water stress, and local environmental conditions – quality and intensity of light, CO_2 concentration, and surface temperature of leaves. We must also remember that transpiration has an important cooling effect on the plant. If all conditions remain constant except the evaporative demand of the atmosphere, the critical value of turgor pressure starts to decrease due to the closing of the stomatal openings. Owing to this increasing stomatal resistance, the rate of transpiration is reduced and causes the actual transpiration to be less than the potential transpiration. With further extraction of soil water, the unsaturated hydraulic conductivity decreases substantially and the compensating increased soil water potential gradient causes a drastic decrease of plant water potential. This decrease is accompanied by a decrease of turgor that closes virtually all stomata. Nevertheless, water vapor still continues to be lost through the plant leaves but at an extremely reduced rate. The greatly reduced transpiration rate allows the cell turgor in the leaves to decrease to such an extent that wilting is clearly evident; see the range of

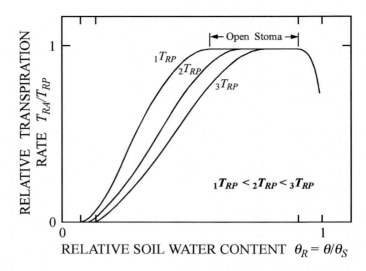

Fig. 10.13 The dependence of the ratio of actual to potential transpiration rate T_{RA}/T_{RP} plotted against the relative soil water content (actual water content divided by saturated soil water content). The curve denoted by $_1T_{RP}$ shows the conditions at very low potential transpiration, while $_3T_{RP}$ represents conditions at very high transpiration. And $_2T_{RP}$ is the potential transpiration roughly in the middle of two extremes

wilting in Fig. 10.13. It is worth mentioning that at soil water contents very close to saturation the actual transpiration can also sink below the potential value. During such times, oxidation-reduction conditions in the soil reduce the root activity of cultivated plants and transpiration.

Up to now we have dealt only with average values of transpiration. However, as was already mentioned earlier in the section on evaporation, the evaporative demand of the atmosphere manifests large diurnal amplitudes. Such fluctuations of the atmosphere occurring during the day and night cause important fluctuations of leaf water potential; see Fig. 10.14. Compared to those of the leaf, the amplitudes of the water potential in the root are smaller, occur later, and manifest a lag or phase shift. In the root zone, diurnal amplitudes of soil water potential driven by atmospheric demand are negligible, except perhaps at shallow soil depths where large daily temperature fluctuations occur.

10.3.3   Wilting Point

We have already shown that if the rate of soil water extraction by plant roots is insufficient to meet a transpiration rate reduced to its minimum, the plant is wilting. A deeper knowledge about plant physiology and transpiration will allow us now to better understand how wilting point was derived. If the plant does not regenerate its turgor and biological activity after a rewetting of the soil, we define such soil water

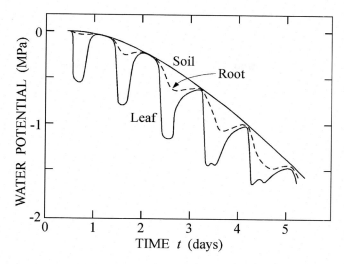

Fig. 10.14 Diurnal changes in leaf and root water potential are distinct, while the soil water potential of the root zone is smoothly sinking during rainless summer days. The changes in leaf and root water potentials are responsible for the diurnal variation in transpiration

content as the soil water content of permanent wilting. Resuming the discussion of the flow of water within plants, we recognize that the soil water content of wilting cannot be the same for different plants. Even if we restrict our consideration to a particular cultivar, the soil water content of wilting depends upon the developmental stage of the plant, its water stress susceptibility, and the evaporative conditions of the atmosphere. Moreover, the composition and concentration of the soil solution plays a role. An increased soil solution concentration increases the soil water content of wilting through a decrease of the total soil water potential. The increased value of hydraulic conductivity owing to an increased soil solution concentration may also influence the value of the soil water content of wilting provided that there are no other concentration effects linked to soil structure and transport across cell membranes.

Because the soil water content of permanent wilting separates soil water into two categories (one useful and available to plants and one that is not), practitioners introduced the concept of the "wilting point" especially for the production of cultivated crops. The "point" is indeed nothing more than an average of soil water content values measured at various times when plants on a given soil were observed to be permanently wilting.

Wilting point is now defined in soil science as the soil water content at a soil water potential of −1.5 MPa (also expressed as −15 bar, −15,000 cm, or a pF of 4.18 which is the common logarithm of the absolute value of the soil water pressure head expressed in cm). The permanent wilting for individual plants and atmospheric conditions usually differ, since the abovementioned value is the result of convention and should be understood in a way as the average of measured data on permanent wilting in thousands of experiments. In this way it is frequently used in water balance studies, in economy models of usefulness of irrigation, etc.

Inasmuch as the wilting point is usually reached when the soil becomes progressively drier, its value is related to the pressure head on the drainage branch of the soil water retention curve. The pressure plate method is commonly used to measure soil water content at −1.5 MPa. Because soil structure does not significantly influence the value, soil samples need not have to be undisturbed in this procedure.

The determination of wilting point based upon biological experiments requires a standardized observation of the permanent wilting of an indicator plant (e.g., barley). Soil texture can be used as an indicator of the wilting point and there exist various empirically based relationships between textural class and wilting point.

Chapter 11
How the Plants Nourish Themselves

We have already mentioned that plants are sort of "natural factories" that have the capacity to produce very complicated organic molecules from a simple form of carbon (carbon dioxide, CO_2). The process of this transformation is called photosynthesis. The term has its root in Greek, *phôs* meaning light, since light is the energy needed to arrange carbon atoms to enter manifold bonds in organic molecules. This conserved energy is released when we burn wood, coal, or dried grass, and everybody has the opportunity to feel some of this released energy as warm air around the flames.

The plants, however, are not satisfied by accepting solely carbon dioxide and dissociated water during photosynthesis. They require many other additional elements known as plant nutrients that are indispensable for their life and growth. These required elements, also denoted as biogenic elements, are easily and simply identified from the chemical analysis of ash after burning the plant's body. The name was derived from the Greek *bios* meaning life and *genos* meaning race or kin and also Latin *genus* meaning descent, family, or type. Autotrophic plants accept carbon from the atmospheric CO_2 that gets into contact with the plant cells in microscopic windows called stomata openings. They can be opened or half closed or completely closed like real windows, as we have mentioned in Sect. 10.3.2. Photosynthesis begins when CO_2 moves into the plant through these windows. The term autotrophic is derived from two Greek words, *autos* meaning self and *trofe* meaning nutrition. A second and somewhat opposite term is heterotrophic that is formed from the Greek *heteros* meaning another. A heterotrophic process occurs during the germination stage of plants' growth from seeds. The emerging plant sprouts are nourished heterotrophically by required matter inside of the seeds up to the time when green organs capable of photosynthesis evolve. With such development, a transitional stage occurs when the plants begin and eventually completely sustain the process of photosynthesis utilizing the inflow of elements pumped into their bodies by their fine roots. Subsequently, even the roots are kept alive heterotrophically. Living without photosynthesis, microorganisms, fungi, and small animals frequently contribute to the heterotrophic nutrition of plants.

© Springer Science+Business Media Dordrecht 2015
M. Kutílek, D.R. Nielsen, *Soil: The Skin of the Planet Earth*,
DOI 10.1007/978-94-017-9789-4_11

There are about 70 biogenic elements in plants bodies. With at least nine of them being present in substantial, easily detected amounts (C, H, O, P, K, N, S, Ca, Mg), they are known as macroelements. Among these important elements, carbon enjoys a special position since it is a part of our Earth's countless natural cycles. Except for C, H, and O, the other six macroelements belong to the products of two basically different processes: weathering of minerals and decomposition of plant residues. They participate in the completion of photosynthetic products with their content within 1 kg of dry plant matter usually being more than 1 g. The remaining biogenic elements occur in concentrations less than 100 mg in 1 kg of dry matter. Because of their tiny amounts found in plants, they are denoted sometimes as trace elements or more precisely as microelements. Among these relatively scarce elements, those manifesting somewhat larger concentrations between 100 and less than 0.1 mg/kg of dry plant matter are Cl, Fe, B, Mn, Zn, Cu, Mo, and Ni, respectively. Such tiny concentrations enable and provide rates of biochemical reactions essential for the production of special types of compounds. The plants accept macro- and microelements dominantly in the form of water solutions through their fine hair-size roots. The acceptance of these elements in the gas phase is not frequent. Fine hair-size roots are short, have lengths of no more than a few millimeters, and live only for several days. They accept the elements as water solutions under the driving force of the gradient of total water potential.

The root system is important for withdrawing nutrients from the soil into the plant. However, an extensive root system does not mean that the plant is more effectively supplied by nutrients – it is merely a prerequisite assuring that nutrients will be absorbed provided that all factors decisive for soil fertility are optimal. On the other hand, under restricted nutrient supplies, a larger root system may allow a more effective "pumping" of nutrients into the plant. Generally, essential nutrient supplies are continually adjusted by specific nutrient demands from the crop.

11.1   Photosynthesis

Photosynthesis is the conversion of energy from the sun into sugars and generally into a plant's body. It is chemical synthesis that is initially conditioned by sunlight. Energy gained by absorption of the sunlight in chlorophyll is used for the synthesis of organic compounds from such simple substances as carbon dioxide and water. Chlorophyll is the green pigment dominant in leaves that is uniquely specialized to absorb light. The gained energy is first utilized in photolysis when water molecules are split into gaseous oxygen and hydrogen ions. The process is accompanied by the release of electrons that move in the electron transport chain that is also well known as the respiratory chain in the final stage of aerobic respiration (Figs. 11.1 and 11.2).

Within the chloroplast there is an enzyme that enables the synthesis of adenosine triphosphate – an adenosine molecule with three inorganic phosphates. Briefly, it is called ATP enzyme. The synthesized ATP preserves and protects energy up to the times when energy is needed. At those times, releases of the needed energy are

Fig. 11.1 The principle of photosynthesis: carbon dioxide + water + light energy = glucose + oxygen. The rate of photosynthesis depends on CO_2 concentration, light intensity, and temperature. The chlorophyll molecules are located on the outside part of the ring-shaped thylakoids that form one of the parts of the chloroplast. In the inside part of the chloroplast, the sugars and starch produced by photosynthesis are kept and protected by the outer and inner membrane of the chloroplast; see the Fig. 11.2

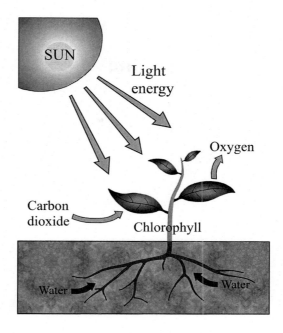

realized by the transformation of ATP to ADP (adenosine diphosphate) by removing one of the phosphates. Such energy is required for a new plant, for the growth of roots, or for the development of new hair roots. Without such energy resources, flowers cannot exist. Another example of energy use is when stomata openings are changed, i.e., when work has to be performed. This ubiquitous ATP molecule of living cells is also used to build complex molecules when it powers virtually every activity of the cell and organisms. At the same time the neighboring cells next to stomata cells have to be informed about this change. Such information transferred by ATP also happens in all instances when a cooperation of cells is required. Without this transfer, the abovementioned changes of stomata closing and opening are not achieved. However, this is only a single example of the generally acting information system. The cycle ATP → ADP and ADP → ATP is repeated in accordance to the external situation. If the release of energy by ATP → ADP was not sufficient, the "release" of phosphate and energy would continue up to AMP (adenosine monophosphate). Here is the source of energy for active transport of ions and molecules across plasma membranes and the synthesis of biomolecules.

Three scientists – John Walker, Paul Boyer, and Jens Skou – provided the basis for a detailed understanding of photosynthesis. They received the Nobel Prize for Chemistry in 1997 for showing how ATP shuttles energy. Walker started the studies by determining the amino acid sequence of the enzyme responsible for ATP synthesis and then elaborated its three-dimensional structure. Boyer showed the reality of ATP enzyme creation in contrast to the previously accepted belief that ATP is formed from synthesis of ADP and phosphate. Skou showed that this enzyme promoted ion transport through membranes and later proved that the phosphate group ripped from ATP binds directly to the enzyme.

CHLOROPLAST

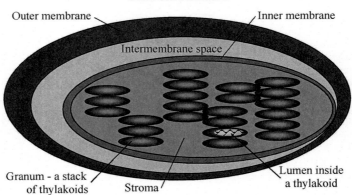

Outer membrane

Inner membrane

Intermembrane space

Granum - a stack
of thylakoids

Stroma

Lumen inside
a thylakoid

THYLAKOIDS

Pigments containing
chlorophyll

ATP synthetase

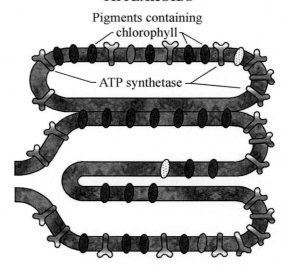

Fig. 11.2 A scheme of the chloroplast and of the thylakoids with pigments containing chlorophyll

In majority of cultivated plants, because photosynthesis involves the conversion of CO_2 into molecules with three carbons, we call this group C3 plants. In some species called C4 plants, the CO_2 is first converted into molecules containing four carbons. If the concentration of CO_2 in the atmosphere is changed, C3 and C4 plants are affected in a different way. In addition to these C3 and C4 plants having different photosynthetic systems, a third group of plants with a completely unusual metabolism of CO_2 also exists. They include 18 families and are called by the abbreviation CAM (crassulacean acid metabolism). Unlike C3 and C4 plants that close their stomata at night, CAM plants keep their stomata open during the darkness of night. Although there are other fundamental differences between C3, C4, and CAM plants, photosynthesis is the basic process for the development of plant organs achieved and completed by macro- and microelements accepted from the soil. In point of fact, even photosynthesis cannot exist without the presence of the element phosphorus accepted from the soil.

11.2 Plants' Pantry

Soil is a pantry for all plants where the following elements are continuously stored and waiting to be accepted by roots as ions of simple inorganic compounds dissolved in water. We segregate the macroelements into primary and secondary nutrients.

The primary nutrients are nitrogen N as ions NO_3^- and NH_4^+, phosphorus P in ionic forms of $H_2PO_4^-$ and HPO_4^{2-}, and potassium K as K^+. The secondary nutrients are calcium Ca as Ca^{2+}, magnesium Mg as Mg^{2+}, and sulfur S as SO_4^{2-}. Among microelements, the most important are the next micronutrients: boron B as $H_2BO_3^-$. chlorine Cl as Cl^-, cobalt Co as Co^{2+}, copper Cu as Cu^{2+}, iron Fe as either Fe^{2+} or Fe^{3+}, manganese Mn as Mn^{2+}, molybdenum Mo as MoO_4^{2-}, and zinc Zn as Zn^{2+}. The division between macro and micro does designate that one nutrient element is more important than another. Instead, they are merely required in different quantities and concentrations. A lack of some of the abovementioned biogenic elements could result in an abnormal evolution of the plant or some of its organs. As a result, the harvest could decrease, and in extreme, the plant could wither and die.

However, soil is sometimes also a medium where toxic elements reside. Among them are Al, Cd, Pb, Cr, and Hg. They initially cause a retarded growth that may well eventually stop the growth of all the plant. They do so by disturbing the function of the stoma which is so important in photosynthesis and transpiration or by disturbing the enzymatic activity and function of plant membranes.

One of the most eminent personalities in studies on plant nutrition was the German chemist Justus von Liebig (1803–1873). In addition to many improvements of chemical analysis procedures, he is famous for being the first person to introduce nitrogen-based fertilizer. Recognizing the possibility of substituting chemical fertilizers for natural fertilizers coming from animal dung, he proposed to supply nitrogen in the form of ammonia to plant roots. Indeed, he initiated the broad use of chemical fertilizers. We recognize now that 30–50 % of crop yields are attributed to natural and synthetic commercial fertilizers. His broad-minded approach is documented by his famous Law of Minimum stating that a plant's development is limited by the essential mineral that is in the relatively shortest supply. Or if one crop nutrient is deficient, plant growth will be poor, even if all other essential nutrients are abundant. It is well demonstrated on "Liebig's barrel," Fig. 11.3.

Fertilization will not compensate for poor soil preparation, the lack of water, weed competition, and other non-nutrient growth-limiting factors. Fertilization will

Fig. 11.3 Liebig's barrel demonstrates his Law of Minimum that declares the element in minimal supply decides the crop growth deficit

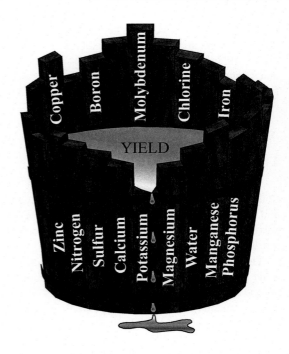

not enhance desired growth if non-nutrient growth-limiting factors are active. From nutritional aspects, the primary difference between manufactured and organic soil fertilizers is the speed at which nutrients become available for plant use. The release of nutrients from manufactured fertilizer usually takes a few days to weeks. Nutrients released from organic fertilizer become available over a period of months or years. Recognition of this difference is critical when farmers use commercial fertilizers.

If the application of commercial fertilizers is high, a lot of fertilizer is washed out by rain or by irrigation water especially with flood or furrow irrigation systems. These excessively leached fertilizers subsequently appearing in rivers, lakes, and sea gulfs support the rapid growth of both microorganisms and macroorganisms like algae. When they die, their bodies are attacked and decomposed by bacteria. As bacteria consume all of the oxygen dissolved in these waters for their activity, a lot of half-decomposed or non-decomposed bodies remain that provide an opportunity for the development of microorganisms adapted to the absence of oxygen. However, the products of their reduction action are toxic for fish and not acceptable for humans. Acidic conditions evolve, the pH decreases, and the acidity of the soil increases. It is understandable that overloading surface waters with fertilizers is ecologically harmful. We are going to describe it in more details in the next chapter at the end of the paragraph dealing with phosphorus. Pollution of groundwaters by fertilizers is less frequent, but if detected, it is a definite signal that the groundwater could be polluted by microorganisms and should not be pumped and used in water supplies.

11.3   Plants' Needs for Individual Nutrients

Nitrogen, N, is frequently a limiting plant nutrient. However, it is not an all-embracing general rule. For example, the vegetative growth of tomatoes, cucumbers, squash, and melons will react to an excess of nitrogen at the expense of fruiting. Trees and shrubs have a relatively low need for soil nitrogen. On the other hand, potatoes, corn, and vegetables like cabbage, broccoli, and cauliflower are heavy feeders and benefit from high soil nitrogen levels. Nitrogen is used by plants in two forms, as *ammonium* (NH_4^+) and *nitrate* (NO_3^-). They are needed for making amino acids from which proteins are subsequently produced.

The principal transformations of N forms are described as nitrification and denitrification. In the first step of nitrification, special oxidizing bacteria use oxygen to transform ammonia into nitrite. In the second step, the nitrite is oxidized into nitrate mainly by *Nitrobacter* bacteria. Nitrification is important in transferring nitrogen into the form dominantly used by plants. Its use is realized if the soil water contains adequate amounts of dissolved oxygen.

Because ammonium is positively charged, it is attracted to negatively charged soil particles, mainly clay minerals (see Sect. 5.2.2). Being less soluble than nitrate, it is more resistant to leaching (movement down through the soil profile). Nitrate, being negatively charged, readily leaches below the root zone with excess rain or irrigation on sandy soils. We prevent water pollution by avoiding overfertilization of nitrogen, particularly on sandy soils.

If oxygen is absent, an opposite process prevails – nitrates are reduced to nitrites, next to ammonia, and then to nitrogen gas. This reduction of nitrates to gaseous nitrogen by microorganisms in a series of biochemical reactions is called denitrification. This process is wasteful because plant-available soil nitrogen is lost to atmosphere. In soils with high organic matter and anaerobic soil conditions (waterlogged or ill drained), the rate of denitrification is even more wasteful.

Denitrification leads to the loss of nitrogen from soil and reduces an essential nutrient for plant growth. In relation to soil fertility and agricultural productivity, it is an undesirable process. On the other hand, denitrification is of major ecological importance in view of the fact that without it, the supply of nitrogen including that in the atmosphere would start to become depleted.

Although soil microorganisms release nitrogen bound in the organic matter of plant residues and decayed roots, the release rates are very slow. Therefore, rates of nitrate availability from this source are not adequate for modern-day crop production that is now practiced in the way prognosed by the German chemist Justus von Liebig. Together with ammonification, nitrification forms a mineralization process that refers to the complete decomposition of organic material with the release of available nitrogen compounds. This decomposition replenishes the nitrogen cycle (Fig. 11.4).

Phosphorus, P, is one of the primary nutrients in plant growth. It occurs in four different aqueous ionic forms surrounded by tetrahedral arrangements of four atoms of oxygen with their eventual one or two or three bonds to hydrogen. The forms and

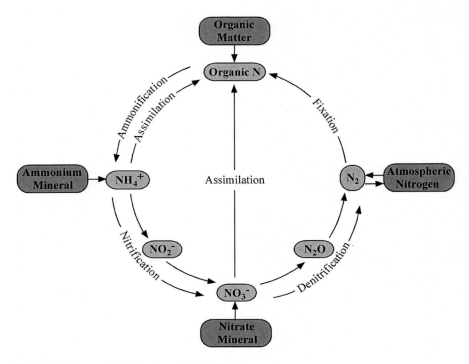

Fig. 11.4 Scheme of nitrification and denitrification

terminology are related to the reaction, and we use the term phosphate with descriptive modifiers. The phosphate ion (PO_4^{3-}) exists in strongly basic conditions. In weakly basic conditions, we find the hydrogen phosphate ion (HPO_4^{2-}). In weakly acid conditions, there is dihydrogen phosphate ion ($H_2PO_4^{2-}$). In strongly acidic conditions, trihydrogen phosphate (H_3PO_4) prevails. See also Fig. 11.5.

Phosphate is important in conserving energy and in taking part in the release and transport of energy within a plant's body, as we have seen in the example on the cycle ATP $\rightarrow$ ADP and ADP $\rightarrow$ ATP, Sect. 11.1. Adequate P availability for plants stimulates plant growth rates especially during their early development as well as hastening their maturity. Phosphorus deficiency is difficult to diagnose because other growth factors give similar symptoms. A laboratory soil test calibrated with tissue tests for specific crops is the best method to determine the need for phosphorus fertilizers.

The phosphate used by living organisms is incorporated into organic compounds. When plant residues are returned to the soil, this organic phosphate is slowly released. A certain portion of it is incorporated into more stable organic forms and becomes part of the soil humus. The release of inorganic phosphate from organic phosphates is called mineralization. The breaking down of organic compounds is mainly realized by microorganisms. The activity of microorganisms is highly influenced by soil temperature and soil moisture. The process is most rapid when soils are warm, moist, and well drained – not waterlogged.

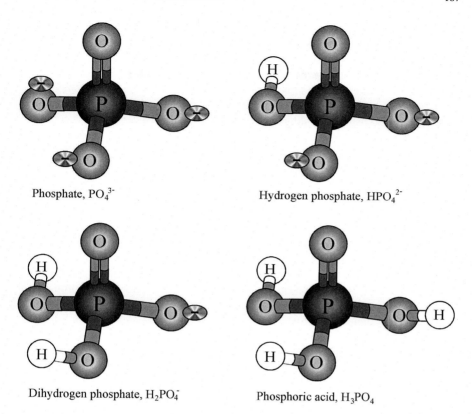

Phosphate, PO_4^{3-}

Hydrogen phosphate, HPO_4^{2-}

Dihydrogen phosphate, $H_2PO_4^-$

Phosphoric acid, H_3PO_4

Fig. 11.5 Models of tetrahedral or quasi-tetrahedral arrangements of P, O, and eventual H in four forms of phosphates

Since the end of the nineteenth century when the great importance of phosphorus was recognized, its deficiency has been compensated by timely applications of industrial fertilizers. Rough estimates report that 30–50 % of crop yields are attributed to natural or synthetic commercial fertilizer. However, excessive applications of phosphorus fertilizers can exacerbate iron and zinc deficiencies and increase soil salinity.

Hence, soil salinization and iron chlorosis could also be the consequences of overfertilization of phosphate. In addition to these undesirable effects, phosphate can be flushed away by rainwater during soil erosion, especially when excessive amounts of phosphate fertilizers are applied. The surface runoff erodes and carries away soil particles filled with an abundance of phosphate that readily accumulates in waters of creeks, rivers, and lakes. When an extreme accumulation of phosphate occurs in waters of rivers and lakes in combination with an excess of nitrates, we speak of eutrophication (from Greek *eutrophos* meaning well nutrified) or hypertrophication (from Greek *hyper* meaning excess or over and *trophe* meaning nourishment). In other words, the growth and decay of algae and plankton dominate the waters of the rivers and lakes. As the algae die, they sink to the bottom where they are decomposed, and the nutrients contained in organic matter are converted into inorganic forms by bacteria. The decomposition process uses and depletes sufficient amounts of oxygen from the deeper waters to kill fish and other aerobic organisms. The water becomes cloudy and murky having typical colors of a dark green, yellow, brown, or red. Eutrophication also decreases the value of rivers, lakes, and estuaries for recreation and fishing. Health problems are especially critical when the contaminated water is used as drinking water for humans because adequate water treatment is difficult to achieve. Up to now we described eutrophication as commonly caused by human activities, but it can also be a natural process particularly in lakes and reservoirs.

Potassium, K, the third major plant and crop nutrient after nitrogen and phosphorus, has been used since antique times as a fertilizer applied to soils as water-soluble salts containing potassium and called potash. The name derives from "potash," which refers to procedures known for obtaining soluble salts from plant ashes soaked in a pot of water. Elemental potassium does not occur in nature due to its violent reaction with water. Potash improves yield, nutrient value, taste, color, texture, and disease resistance of food crops. It has wide application for the production of fruit, all grains, sugar, corn, soybeans, and cotton. General symptoms of potassium deficiency include chlorosis – the yellowing of leaves caused from a lack of chlorophyll – followed by scorching that moves inward from the edges of the leaves. Older leaves, affected first, may crinkle and roll while shoots suffer shortened, zigzag growth. The excessive amounts could act harmfully in semiarid and arid zones because of increased salinization.

The three remaining macronutrients are frequently called *secondary macronutrients*. *Calcium* Ca is an essential part of plant cell walls. Their structure enables transport and retention of all elements constituting the growth of the plant's body. If calcium resides in soluble soil compounds, it acts against the accumulation of alkali salts and balances the tendencies toward the acidification of soils. Hence, its presence in a soil is highly beneficial for the great majority of cultivated plants. Accordingly, liming is a very effective tool against acidification, especially after the drainage of waterlogged soils. As we have shown in Sect. 11.1, *Magnesium* Mg forms the part of chlorophyll in green plants that is necessary for photosynthesis and growth of plants. Consequently, whenever magnesium sources in soils are inadequate, dolomite liming or the application of fertilizers containing Mg is a common agricultural practice. *Sulfur* S is essential for production of proteins in plants and for enzyme activity as, e.g., related to photosynthesis. It improves the growth of roots and seeds and enhances the resistance of plants against rapid changes of meteorological conditions.

Micronutrients act in individual processes and their role differs in relation to individual types of plants. Their many special roles are described primarily in academic literature.

Bioenergy has to be mentioned when we deal with nutrients. Nowadays, a huge advocacy and trend of industrial-scale harvesting of biomass for bioenergy production exist owing to political, pseudoscientific, and short-term economic endorsement. Here we refrain from using the terms biofuels and bioenergy as if oil and coal were not the clean products of biologic processes. Indeed, the production of "biofuels" presents a potential environmental risk. The long-term harvesting of biomass for bioenergy production will have an adverse impact on soil quality, a loss of soil structure with consequent worsening of soil hydrologic conditions, an increase of flood frequencies, and a decrease of agricultural productivity. The extensive production of "biofuels" causes a loss of readily mineralized organic N and of other nutrients. The loss of cation exchange capacity due to the decrease of humus content indicates that long-term residue harvesting reduces the ability of the soil to retain cationic nutrients. Without compensating organic amendments or other conservation strategies, the long-term harvesting of crop residues results in degradation of the quality of soils.

Chapter 12
Soil Reveals Its True Character

12.1 What the Soil's Face Says

"It's a nonsense," emphatically stated by the serious soil scientist who was famous for multitudes of papers published under his name as the first author. And as if those were not enough, there were twice as many additional papers where his name appeared among the names of coauthors. Jealous gossips used to say that everybody within this famous scientist's institute was obliged to include his name as a coauthor of any of their manuscripts.

"It's a nonsense to offer a face to a soil. Although I am aware of the repetitious expressions of fundamental pedologists that soil is a living body, it is only the medium for life. Soil can be favorable, but in some instances it is less favorable or even unfavorable for living organisms," the famous soil scientist explained to both of us.

We readily agreed with him. Our concurrence was not just politeness nor was it our ductility owing to our substantially lower number of publications. It is our standpoint from which we have evaluated all of our research. This entire book is our witness.

We started our discussion with the famous, yet skeptical, soil scientist by arguments eventually supporting our sentence about soil's face. Here, we explain why we use the word face. Ages ago, people were grouped according to features of the face. If somebody had great irregularities in his face, he was considered as dangerous for his pack or tribe, and when the tribe started to be called nation, the man with a disfigured face was believed to be some sort of disaster or danger for the nation. This belief was probably the reason why the face of criminals was frequently disfigured in the prison, even though it may have happened in a fight before they were imprisoned. In those days, scoundrels released from jail hit on a good ploy by keeping the scars hidden under their beard. Nowadays, they pay for plastic surgery. Or from just another corner, men developing their bodybuilding today believe that pretty women will prefer them provided that they keep a proud and very masculine

© Springer Science+Business Media Dordrecht 2015
M. Kutílek, D.R. Nielsen, *Soil: The Skin of the Planet Earth*,
DOI 10.1007/978-94-017-9789-4_12

face. Moreover, pretty women should look like a cast of a famous lady, an actress, or a feministic VIP. But there is a principal "but." Some men look like cowards, but everybody knows that roughly the same percentage of real cowards is found among brave-looking men. As far as the majority of us believe, these and other facts contributed to the discovery of our many recent objective tests. Hence, today, individuals believe and rely completely on the results of blood-related tests to detect genetic relations, e.g., the kinship and kinfolk. And how is it with our soils? What is important on their faces? Which features are more important than others? How does each of them differ from the others?

We were so deeply submerged in our own argumentation that we did not notice that our famous soil scientist left us, probably to sign another paper. We are now left alone, but we enthusiastically continue to explain what we consider as the face of the soil.

First, soils have their own typical and characteristic features and sizes, i.e., they have an assortment of features at specific depths. In regions of mild climate, a soil depth may extend down to 150 cm, but for a tropical climate, a depth of several meters is nothing extraordinary. On the other hand, because of unfavorable factors of soil evolution, shallow soils are very common within specific continental regions. An excellent example is the existence of very shallow soils in semideserts caused by a regional semiarid climate. When we try to recognize the principal features of a soil, we often dig a pit into the soil surface having a horizontal size about 0.6 by 1.5 m that reaches a depth of 20–30 cm below the parent rock (material). Quite frequently, we also recognize soils on the surface of continental landscapes that were developed from river sediments (alluvium) or from dust (loess) blown into regions by strong winds during earlier glaciations that lasted for about hundred thousand years during the last half million years. Neither of those materials has the appearance of a rock, but we still use the words *parent rock* for all inorganic materials transformed to soil at the interface with the atmosphere.

Even a layman will recognize at least two "layers" when he carefully examines the vertical wall of a pit dug into the soil surface. We prefer to call them *horizons*. The name has its roots in the real, truly horizontal arrangements of these relatively thin "layers." They are neither inclined nor folded as it often happens to geologic layers. Quite opposite, they are like giant thin pancakes placed flatly one upon the other by Goliath and monotonously extend horizontally in all directions from our observation pit. This horizontal arrangement is the reason to call them horizons. This terminology has prevailed more than one and a half centuries.

Each soil horizon is characterized by an accumulation of a clearly identifiable family of materials or by the loss of some members mainly due to the dominant direction of flux of water. Another distinctive sign for a soil horizon is the characteristic transformation of matter by biochemical processes. When we stand in our freshly dug pit, it is usually the first time that we come face-to-face with this soil. We try to recognize the horizons according to the color and hue. We classify the type of the structure. Are the aggregates rounded or do they appear more like cubes or polyhedra? Or are they platy or like flakes? How are they separated from another? When we wet a small clod in the palm of our hand with water and rub this wetted material between our fingers, we can roughly judge if the soil is sandy, loamy, or clayey, i.e., the soil texture. We focus our attention upon the potential differences in texture between the horizons, especially between those horizons lying below the top humus horizon. Further important features for identifying individual horizons are the density of fine roots including the depth of their penetration, the origin of a specific neoplasm, and whether or not the transition from one horizon to the next is sharp or gradual. The word neoplasm, derived from the classical Greek *neo* meaning new and *plasma* meaning formation, denotes newly formed bodies originating from the products of weathering and humification that are by-products of soil-forming processes. After the soil horizons forming the entire soil profile visualized in the excavated pit are described, soil samples taken from each horizon are quantitatively analyzed in the laboratory. The chemical and mineralogical content as well as the soil physical and soil hydraulic properties of each sample are quantitatively ascertained. All these data coupled with microscopic observations of soil micromorphology offer detailed keys to appreciate and understand how mineral particles are arranged and interconnected by amorphous and humified materials whenever we observe thin, undisturbed soil sections in a microscope. Combining the field description of individual horizons composing the soil profile with results of laboratory tests enables us to identify the soil taxonomic type.

The birth of soil – or soil genesis in scientific jargon – changes the parent material from which the soil originated, and this transformation is generally recognized even by the naked eye of a layman and confirmed by the soil scientist. For example, a soil originated from a loessial sediment is evident to everybody, but a more detailed evaluation of its soil genesis leading to its taxonomic classification is more complicated and should start with the identification of its soil horizons.

12.2 Soil Horizons

Soil horizons are defined as diagnostic soil horizons when they reflect features that are recognized as occurring in soils and can be used to describe and define soil orders, classes, families, or types.

Inasmuch as the horizon at the soil surface is characterized by accumulations of humus, it is named the humus horizon. Its dark color that can even be black stems from the color of humic substances, especially when they form a sort of film on solid silicate particles in addition to humus filling a small part of its soil pores. The international designation is A horizon that has its roots in the original system of horizons used by the classics of soil science. At that time the first letters of the alphabet were used, and since the humus horizon was the first when we start the description from the top, it got the abbreviation A horizon. The next horizon if present got the abbreviated name B horizon. Below it where parent rock exists is denoted as C horizon. Sometimes the C horizon is just below A horizon. This alphabetic terminology for soil horizons was adopted without any modification for about a half century until O horizon was introduced to designate more or less decomposed

organic debris on top of a mineral A horizon without mixing with its mineral particles. But let us return back to the origin of the description of horizons. With time
and with detailed field studies, new horizons and sub-horizons denoted with small
letters taken from the process dominating their formation were added to the basic A,
B, C symbols (e.g., Ap-Bv-C profile of the soil type cambisol). Here p means
plowed, and v is for the German Verwitterung (weathering). In this way a sort of
pedological stenography was developed. In some instances we can find another
horizon between the A and B horizons, and then the sequence is A-E-B-C in our
pedological shorthand. More about this eluviated E horizon will be discussed later
on after we describe the processes in B horizon formation.

When clay and silt particles have been transported from the A horizon to deeper
positions during long time periods of a thousand or more years, a horizon rich in
those fine particles is formed below the A horizon. It is called an illuvial horizon
derived from the Latin *lavere or lavare* meaning to wash and *inlavere* later transformed to *illuvere* meaning to wash in, to depose material washed out from the top,
or generally from somewhere else. The letter "n" was transformed into "l" during
the ancient Roman times just for ease of pronunciation. Using the pedological shorthand of the end of the nineteenth century, this illuvial horizon is now recognized as
the B horizon. The main factor causing change within this B horizon has been the
climate during those years when the annual precipitation is higher than the annual
evaporation measured from the free water level of a meteorological station. Under
this climatic condition, it was typical that water flowing down in the soil prevailed
over that flowing up with the consequence of clay particles being transported and
deposited in the B horizon. Local soil particles were covered by a thin film formed
by clay minerals with a typical orientation, and polyhedric structures were formed.
During their transport, since the clay and fine silt particles are frequently covered by
iron oxides originating due to chemical weathering, the B horizon has an intensive
brown color. If the soil is acidic with pH well below 6.5, the intensive weathering of
primary minerals releases free simple forms of Fe and Mn oxides that contribute to
the clay film cover on the surfaces of soil particles. The color becomes more reddish
and rusty. Soil scientists in their detailed studies have recognized differences
between horizons described as, e.g., Bv, Bt, and Bhs where the small letters accentuate the specific features of the horizon-forming processes. Having already
explained that "v" is for intensive weathering (German *Verwitterung*), "t" is for
German *Ton* meaning clay indicating that the horizon or sub-horizon is enriched by
clay particles compared with their content in neighboring horizons. The symbol
"hs" designates increased accumulation of humus and sesquioxides according to
chemical analysis.

In order to avoid any misunderstanding regarding the evolution of a B horizon,
the mutual arrangement of soil particles must also be considered. Coagulation and
microaggregation act against illuviation because they mechanically hinder the
transport of fine individual particles through soil pores.

Generally, the accumulation of matter in B horizons leads to the possibility of
waterlogging during periods of wet weather that leads to a lack of oxygen in chemical processes and the development of reduction processes. Because reduction

processes are typified by small patches of rust and pale yellow colors, pedologists speak about marmoreal sub-horizons. With those sub-horizons being less permeable than their neighboring sub-horizons, they contribute to more intensive waterlogging and increased acidic conditions. These lower pH reactions repeatedly intensify reduction processes. This repetition stops in dry seasons with low precipitation. Or more precisely speaking, when the oxidation starts, the Fe^{2+} and Mn^{2+} compounds originally dissolved in water are transformed by oxidation into chemical compounds not soluble in water. These compounds precipitate first and are more distinctly visible in coarse pores because they drain first and the oxidation is more intensive. Therefore, the visibility of marble colors is more distinct.

Soil horizons rich in sesquioxides are formed by different solubility in water illustrated here with iron. Iron in the ferric form (with valence 3, i.e., Fe^{3+}) is relatively immobile. If soil drainage is poor and most or all soil pores are filled with water, the minuscule amounts of soil air are rapidly used by microorganisms which could not exist without oxygen. They are then forced to obtain oxygen from chemical transformation of ferric (trivalent) forms into ferrous (divalent) forms. Ferrous iron (FeO) dissolved in soil water moves and migrates in the soil while kept in the reduced form, i.e., during and after rainy periods whenever the soil remains water saturated. During the whole period of waterlogging, the soil iron is free to migrate locally within the horizon. Whenever the rainless period lasts sufficiently long to allow air penetration and oxidation reactions, the soil iron is transformed into water-insoluble ferric forms. Precipitates of such ferric compounds are visibly identified as mottles. Amplified lengths of such periods lead to intensification of the process and to more frequent mottling. In tropical climates the process is so strong that it dominates the formation of thick horizons rich in iron and accounts for the birth of latosols (lateritic soils). The name, derived from the Latin *later* meaning brick, relates to the intensive red color of these soils that is similar to the color of bricks produced of loess and baked at high temperature.

In cool regions with high values of annual precipitation, horizons poor in Fe compounds are formed by leaching sesquioxides during wet periods when intensive downward fluxes of soil water prevail. Such a leached out horizon is evident by its gray whitish color similar to ash, Russian *zola*. Together with Russian *pod* meaning under or possibly with Ukrainian farmers' slang *poda* meaning soil, the name podzol was commonly used in the countryside and eventually entered into soil science literature. Podzolic horizons are formed overlaying a sesquioxide-rich horizon. The leached or eluviated horizon was earlier denoted by A2, nowadays by E. The term eluviate stems from the Latin *exlavere* which is a combination of *ex* meaning out of and *lavere* meaning wash. Analogous to *illuvere*, the term *eluvere* was created. Today, every indication of increased or decreased content of sesquioxides is easily recognized in the field, and the more distinct is the change of color in the direction to rusty red, the more intensive is the process of horizon formation.

We have briefly described the most frequently occurring main horizons and processes leading to their formation. However, the spectrum of variation of horizons is much broader and their characteristics go into many details. For example, the World Reference Base (WRB) supported by the International Soil Science Society and

FAO lists 40 diagnostic horizons. The US Soil Taxonomy distinguishes as many as 30 surface horizons (epipedon) and subsurface horizons.

When we summarize this chapter on soil horizons, we have to accentuate that individual horizons differ one from another by characteristics readily visible while looking at soil profiles in the field. They are color, structure, texture, resistance to rupture, and deformation. We can also detect the presence or absence of carbonates and obtain reliable estimates of pH. The field detection of soil morphology is then completed by laboratory analysis. But the first step in our study of the soil is in the field where we describe the soil profile and all observed features (Fig. 12.1).

Fig. 12.1 Studies of the soil in the soil pit. Soil faces differ from one location to another and are not uniform. We show the variability of some of them. The soil profiles demonstrated here reach roughly to a depth 1 m below the surface. The type of each soil is in the US classification at the level of order and suborder. In the World Reference Base, WRB, the Reference Soil Group, RSG, is indicated

12.3   A Long and Prickly Start of Predecessors of Soil Science

We can find the first notes about soils already at the start of written documents of great civilizations. It is without doubt that the people had a certain knowledge about quality of soils at the stage of nomad lifestyle and that the great revolution of the beginning of agriculture was closely related to understanding which soils would bring the best "harvests." After some members of a tribe practicing gathering and hunting had accidentally and just by chance spilled a handful of seeds, they later "harvested" there more than just from the natural vegetation in an alluvial soil near a river. When similar incidences were repeating and happening for several years, the nomads started to gain basic knowledge about seeding, providing that it happened on fertile soil. The importance of that provision during times when the greatest revolution in mankind was rooting cannot be overestimated today nor fully appreciated. They "harvested" much more than in the practice of gathering, and they were less menaced by famine again and again provided that they recognized fertile soils from those of low fertility. Their successful recognition of fertile soils is indeed only our speculation derived from the archeological discoveries at Abu Hureyra in today's Syria on the bank of Euphrates near the blind meander of the river formed between 13,000 and 9,500 years ago and generally in the Fertile Crescent. We are not dealing with the various factors accelerating the start of farming such as climate instability that allowed human survival under conditions of planting the first sorts of hay and barley. About 1,000–2,000 years later, other centers of primitive farming appeared independent of Fertile Crescent under different environmental conditions in Yangtze and Yellow River basins, again on fertile soils. Here we wish to concentrate only on the simple, intuitive recognition of fertile soils in Neolithic times.

The simple roots of agriculture appeared about 14,000–13,000 years ago when the glacial (called Wisconsin in the USA, Würm or Weichselian in Europe, etc.) was approaching its end. It could be even stated that some signs of the glacial disappearance occurred 1,000–2,000 years earlier. Because the warming did not show the same intensity in various regions, the same is valid for the start of agriculture. The warming had not the same intensity in various regions and the same is valid for its start. In today's mild zone people used the benefit of increased quantities of foodstuffs – both plants and game – as a result of climate warming. Desert zones were losing their extreme aridity typical for glacial periods, and during the warming transition, the rains arrived regularly each year. Specialized plants together with sheep and followed by goats and probably chickens were coming back. They were arriving together with beasts of pray like big cats.

The transition into our recent interglacial (the period between two glacials) called Holocene was not smooth at all even in the same region. It looked like a climatic stuttering. The climate change had different intensities and lengths in various regions. There were two stages when extreme cooling interrupted the gradual warming, as if the hard climate of glacial was coming back. These two returns are called *Dryas*. The Older *Dryas* was the stadial, or cold period between the two warmer phases about 14,000 years ago, and was not observed in some regions. The first embryo of agriculture appeared in Older *Dryas* when the tribe experimenting with seeds had more chances to survive than the classical hunters and gatherers. The embryonic news spread in Fertile Crescent, and the simple agricultural discoveries were kept at least partly alive during the next warming stage. They were "discovered" again as an efficient tool in the fight against the much harder hit of cooling in Younger *Dryas* (12,800–11,700 years ago). It seems entirely natural that these first steps in discovering advantages of agriculture were practiced on more fertile soils. The climate was characterized by long dry seasons in today's mild zone. The dryness favored annual plants which die off in the long-lasting dry season leaving dormant seeds for the new birth in the next year. These plants put more energy into production of seeds than into strong body and woody growth. The abundance of wild grains contributed to the extension of the starting agriculture, all happening within 1,000 years or more. We must conclude that agriculture is characterized as a coevolutionary adaptation of *Homo* and plants under the provision of recognizing the fertile soil. Accompanied by wildlife, this coevolutionary process was started with the domestication of plants on fertile soils within favorable environments and completed with the gradual domestication of suitable animals.

Thus far, all of our statements in Sect. 12.3 are mainly hypothetical regarding the appreciation of soil quality. Our statements lack direct confirmation and could be rejected on the basis of more facts from solid archeological findings. We know that even the hunters and gatherers were not aimlessly wandering across the countryside. Members of a big family or a small tribe were occupying an area up to about 100 km^2 within which they were moving according to an inherited system, either visiting the places in shape of arrows directed from one center or wandering in a spiral. They were protecting their own area from wild incursions, since they knew well the advantages of the soil they protected from wild nomads. With the start of agriculture, those earlier-mentioned centers were gradually transformed into permanent settlements, and the protection of fertile soils was transformed into organized systems. Here, we have been describing the situation on our Fertile Crescent. But similar archeological findings were discovered in Chinese Diatonghuan and Xianrendong, and due to the influence of Younger *Dryas*, the original nomads gradually changed their style to settlement after first finding the best fitting soils for rye domestication.

But let us shift from indirect documents to real proofs formulated upon the principle *litera scripta manet verbum ut inane perit*: "The written letter remains, as the empty *word* perishes." Here we replace *word* by *hypothesis*. We start with hieroglyphs written in old Egypt. The fertile soils in the valley of the river Nile originated on alluvial deposits of the river. The soils were regularly "fertilized" each year by the Nile floods rich in suspended soil particles and dissolved products of weathering mainly from the upper part of Blue Nile. The name *kemet (kmt)* of those soils was translated as black land. The importance of the *kemet* soil is obvious from the fact that the name of old Egypt starting from pre-pharaonic past or the predynastic period was identical with the name for the Nile's fertile valley soils. In contrast, *deshret* was the term for the reddish desert soil surrounding *kemet. Deshret* was the seat of evil spirits and death. By the way, the modern term desert has common roots with *deshret*. Later on, but still in ancient Egypt, the soils were classified according to their quality, and they had commercial values. The cost of the best of them, *nemhura* soils, was three times more than *shata-teni* soils. There are no known indications about soil quality in Mesopotamian culture except for the differentiation between soils suffering from strong salinization with a distinct decrease of harvests and slightly saline soils where the harvests were not yet influenced by salinization.

In the fourth century BC, the Greek historian and philosopher Xenophon wrote about the system of two soils in his treatise *Oikonomikos* (more familiar under the title *Oeconomicus*) where he devoted a component to agriculture. Translated in our recent language of soil science, we speak about his two fallow-field systems. One field is used agronomically, and the next field is land lying fallow. After 2 years the fallow is plowed and seeded while the first field is left fallowed. The two-year period was not strictly kept: it was frequently exceeded. The famous Roman politician, orator, and writer Marcus Tullius Cicero considered the Xenophon's *Oeconomicus* three centuries later so important that he translated it into Latin. One century earlier the first great Latin prose writer Marcus Porcius Cato (Cato the Elder) wrote a book about agriculture having the title *De Agri Cultura*. The book is wise but also amusing. In the first chapter that deals with establishing and equipping a farm, he wrote that the price of land depends upon the quality of soils. He recognized 21 classes of soils mainly according to fitness of each of them for the highest harvest of a certain plant and ultimately of a small group of plants. In practical parts of the book, he recommends the best time for performing a certain activity on specified types of soil. As, for example, he recommends the best time for tillage. In the spring after a wet winter, the farmer should start plowing sandy soils first and, then later on, continue to plow the remaining soils. Cato is an important author for us soil scientists owing to his perceptive observations of nature followed by his practical recommendations in farming. When his famous phrase *Ceterum autem censeo Carthaginem esse delendam* (English: "Furthermore, I consider that Carthage must be destroyed") was finally realized and the Romans razed the city of Carthago in the third Punic War (149–146 BC), the fertile soils of Carthago's land and colonies partly compensated the Roman soils already attacked and even destroyed in some regions by water erosion, again according to Cato writings. Hence, we appreciate and know that Cato's famous sentence as he used it at the end of each of his speeches had something in common with soil, too.

Following orators' tradition we should repeat a slightly modified sentence *Ceterum autem censemus, solum providerum esse* (Furthermore we consider that we must take care of the soil). There was a substantial difference between classical Greeks and classical Romans. Compared to Greeks who pursued arts and philosophy, the Romans commonly devoted most of their time to practical aspects of life and ensuing activities. Hence, we find unique essays and books about agriculture and the importance of soils written in Latin. One century after the writings of Cato, another Roman writer and politician, Marcus Terentius Varro, presented the treatise *De Re Rustica* (About Agriculture). He was first engaged in political activities, e.g., he was the member of the commission of 20 that carried out the great agrarian scheme of Caesar for the resettlement of southern provinces Capua and Campania. However, he did not reach high political positions. On the contrary, he was in later actions unsuccessful, he was accused of misappropriation on Sicily, and finally he was lucky to gain Augustus' protection so that he could devote himself to studies and writing. He was a highly productive writer and turned out more than 74 Latin works on a variety of topics. It has been traditionally repeated that Varro had read so much that it is difficult to understand when he found time to write, while on the other hand he wrote so much that one can scarcely read all his books. Varro recognized about 100 soils and the criterion for defining each one was the yield of specific plant.

One of the most important writers on agriculture in the Roman Empire was Lucius Junius Moderatus Columella. With advice from his uncle and decades of his own farming experience, he benefited by combining his practical knowledge with studies of his predecessors Cato and Varro. However, many times he quotes the poet Virgil (Publius Vergilius Maro) – mainly his poem entitled Georgics (derived from Greek, "On Working the Earth"). The structure of the completely preserved *De Re Rustica* (About Agriculture) of 12 volumes contains one separate volume on soils. Columella proceeded to a more substantial description of soils, as, e.g., "…the soil and the climate of Italy and of Africa, being of a different nature, cannot produce the same results." Using modern-day language, the first volume contains the principle of soil protection from erosion and advocates that all organic leftovers from agricultural production as well as sewage from herds of cattle should be brought back to the soils. Next, we are quoting the translation on today's website belonging to Bill Thayer in order to illustrate the style how the understanding the soil nature is combined with practical procedures: "There should also be two manure-pits, one to receive the fresh dung and keep it for a year, and a second from which the old is hauled; but both of them should be built shelving with a gentle slope, in the manner of fish-ponds, and built up and packed hard with earth so as not to let the moisture drain away. For it is most important that manure shall retain its strength with no drying out of its moisture and that it be soaked constantly with liquids, so that any seeds of bramble or grass that are mixed in the straw or chaff shall decay, and not be carried out to the field to fill the crops with weeds. And it is for this reason that experienced farmers, when they carry out any refuse from folds and stables, throw over it a covering of brush and do not allow it to dry out or be burned by the beating of the sun."

All Roman writings have a common feature: they were factually written using simple Latin terminology in the unvarnished style of technological manuals. Greeks, on the other hand, wrote about agriculture in a lofty poetic style. With increased time Roman knowledge of soil became much more close to scientific and objective observations of nature's environment. Romans understood that plants took some important substances from the soil and that those substances remain in plant remnants as well in wastes of animals that are fed by plants. They were aware that these substances participated in some sort of a cycle that should not be completely disturbed. Knowing that soil was a natural object, Columella asserted that its profile was composed of two horizons – the top dark one and the lower subsoil. With the fight to eliminate weeds being one of the main problems of agriculture, various types of tillage were gradually developed and recommended by Columella. In order to slow down a decrease of fertility or even to stop it, the liming of soil with ground marble was mentioned in Columella's *De Re Rustica*. Noticing its simplistic name in English, *About Agriculture*, we learn that the antique Romans did not get a head- ache inventing titles of their practically oriented books. The content was much more important. Accordingly, we should not be surprised that Columella devoted several paragraphs to soil erosion and its consequences.

Lasting for centuries after the time of classical Romans, the quality of soils was guessed just for estimating taxes. In Europe there were instructions for deciding to which of three bounty classes a soil belonged. By the end of the seventeenth cen- tury, we find that the system already consisted of eight bounty classes determined by the size of harvest. In spite of Renaissance love for Greek and Latin writing, it is apparent that the first observations of soil properties described by Varro and Columella were forgotten.

Parallel to Europe, there were several other centers of development of agriculture with their own approach to a simple classification of soils. As for example are the Aztecs who divided soils according to their utility and who recognized soils with agricultural potential and those useful for other practical applications such as pot- tery clay. The Indus Valley Civilization developed on the fertile plains of the river Indus. Soils were probably classified into three separate types according to their productivity: those where the fertility was regularly kept at high levels by sequences of river floods, where such renewal of fertility was absent, or where the threat of salinity existed.

12.4 Pedological Darwin

A completely new approach to observe and subsequently study soil could start only after a transition from alchemy to chemistry was successfully realized. After that transition was completed and coupled with the first steps of mathematical descrip- tions of physical laws, a more comprehensive understanding of soils and nature began. Up to 80 years of the nineteenth century, the top part of the soil was studied with research ending at the bottom boundary of the root zone of plants. The research was mainly focused to determine the amount of plant nutrients in the A horizon. The most famous was the German chemist Justus Liebig (1803–1873). He was pro- moted into the rank of noblemen as Justus von Liebig (1845). By this promotion, his extraordinary merits in development of sciences were publicly appreciated. He pro- vided a revolutionary concept for studying soil, since he was the first to introduce newly developing chemistry to soil-plant relations.

Two crucial factors account for Liebig's decision to become a chemist and have a major interest in agronomy, especially in plants-soil relationships: (1) Liebig, apprenticed to the pharmacist practice of his father, performed his first experiments in his father's pharmacy. (2) At the age of 13, Liebig lived through the year "without a summer" when the majority of food crops in the Northern Hemisphere were destroyed by a volcanic winter caused by a large cloud cover composed of volcanic dust, lack of sunshine, and the global famine that ensued. Thanks partly to Liebig's introduction of fertilizers in agriculture, the 1816 famine was called the last food crisis in the Western world.

Liebig's life in sciences did not start in a simple way. He attended the University of Erlangen, but due to his involvement with a radical student organization and due to his criticism that not enough advanced development of chemical studies was available at the university, he left Erlangen without gaining the doctor title. He obtained it later, was appointed a full professor at Giessen University at the early age of 21, and soon after developed a famous, unique chemical center. After World War II the university was officially renamed after him. He founded and edited from 1832 the leading journal in chemistry, Annalen der Chemie, which was renamed to Justus Liebig's Annalen der Chemie after his death. In this journal he also published about the role of humus in plant nutrition and discovered the important role of nitrogen compounds in the feeding of plants. One of his most recognized accomplishments was the invention of nitrogen-based fertilizers substituting for "natural" sources of nitrogen like plant remnants, manure, animal dung, etc. Many of his actions as well as his theoretical manner of proceeding formulated before experimentation can be characterized by his sentences: "…the production of all organic substances no longer belongs just to the organism. It must be viewed as not only probable but as certain that we shall produce them in our laboratories."

He was frequently named as father of production of fertilizers and their application in modern agronomy. When he studied plant nutrients, he formulated the Law of Minimum that a plant's development is limited by the element that is in a relatively shortest supply. The law is simply named "Liebig's barrel"; see also our Sect. 11.2 and Fig. 11.3. This Liebig's Law of Minimum was later extended by ecologists to all forms of life outside the vegetable kingdom. The law was later also criticized for its static character and that it does not consider the dynamic system in which the lack of one factor or element could be partly and in certain limits compensated by the action of another factor.

Liebig's influence upon soil science research was enormous with soil chemistry being studied according to his ideas for several decades leading into the twentieth century. Toward the end of that time span, many universities and research institutes began putting an equal sign between soil science and soil chemistry of plant nutrients. Their promotion of this overly simplified equality, never accepted by Liebig, soon started to break down his inspirational message for others to continue and expand their scientific explorations to better understand all of the intricate concepts and processes within soils.

During the same time of Liebig's dominance, geologists were also identifying various kinds of soils, but their concept was limited to the description of soils as weathering products that were later transported either by wind as, for example, the loess, or by water when alluvial soils developed. Examples of many other types include diluvial deposits on the bottom part of slopes or old moraines after the previous existence of glaciers. All those results of geological research are correct, but the real development of soil was ignored, probably because the thickness of soil was often an order of magnitude smaller than the thickness of transported and deposited materials. During the second half of the nineteenth century, owing to the fact that initial observations were not classified and documented as they are now in a rational scientific manner, little progress was achieved to better understand the origin of soils.

We don't intend to say that there were not written reports where authors tried to study soils to eventually understand their origin. We have mentioned already in Sect 4.6 that Darwin studied the role of earthworms during soil formation by mixing mineral substances with products of humification. He described his experiments in a special publication in 1881. At that time he judged the quality of his soil formation research to be higher than that of his 1859 formulation on the origin of species. The start of his interest in soil formation was very simple. When his son was running on the small hillside of Darwin's land, pieces of very thin rock chips were creaking under his shoes. Thirty years later when Darwin walked and ran on the same small hillside, there was no creaking. Wondering why, he decided to have a ditch excavated across the small field and learned that the creaking was no longer possible because the skeletal particles had completely disappeared without a trace. Again wondering why, he had the length of the ditch sufficiently increased in order to observe natural occurring variations within soil profiles along the entire extent of his property. From his far-reaching observations and the disappearance of skeletal particles, he concluded that the soil was evolving and that earthworms had a great influence upon its evolution by mixing decaying remnants of plants with mineral substances that were also simultaneously changing owing to weathering.

Although we could find similar formulations about the evolution of soils published in the nineteenth century, systematic descriptions of soil horizons and classifications of soils according to definite taxonomic rules were still lacking. In spite of the fact that new ideas are in the air in the discoveries of our contemporary sciences, not everybody is methodical and specialized in an entire branch of science. This was what happened to Darwin in pedology, when he accredited evolution to soils. The application of evolution principles to soils was the first step performed by Darwin. But the full and complete description of soils as a new and complex phenomenon with all signs typical for a dynamic evolution leading to the great variability of soil types was the substance of the life achievements of V. V. Dokuchaev.

12.5 Everlasting Stimulus from Dokuchaev

The first publications about evolution of chernozem and then on soil evolutionary taxonomy appeared 2 years later written by V. V. Dokuchaev, the Russian professor of Saint Petersburg University. It was a famous university with excellent professors like the chemist D. I. Mendeleev, the mathematician P. L. Chebyshev, or the physicist H. F. Lenz. Dokuchaev studied geology, mainly the Holocene. While he studied at the university, he earned additional money for his own needs by privately teaching children of rich parents. He participated at the carousels of his rich students, and he liked this style of life without doubts, qualms, and worries. When his supervising university professor asked him how he was proceeding with the doctorate dissertation, Dokuchaev answered truthfully that it was slow because he did not have enough time.

"Why? Do you have financial troubles?" asked the professor, thinking how to help his talented student. Dokuchaev was not only gifted, but also he had good luck. During his early university studies, he discovered a mammoth skeleton in some alluvial sediments while on a university excursion. Even scientific research is difficult to perform without good luck.

Dokuchaev answered the professor's question, "I am attending rave-ups and drinking parties. They don't leave me enough free time for research."

"Give up all your studies. You are lost and unfit for science," said the completely upset old professor.

Dokuchaev did not give up. He worked hard and he had the special gift to observe and to describe what the others missed in one way or another. All of his observations and data collections were based upon solid foundations of the sciences known at that time. He combined stubborn field studies with moments of good luck as well as bad luck, and finally he became famous just by his completely new ideas on how to define and describe soil. His name has been frequently quoted up to now all over the world, while that of his old, antagonizing professor disappeared in the dust of archives.

Dokuchaev's first publications were devoted to field studies about the genesis of riverbeds and of alluvia around them in Central and Northern Russia. Further on, he wrote about the genesis of loessial wind sediments as well as for the first time the discovery of the zonal extension of a particular type of soil – the chernozem. He borrowed the name from the popular Russian chernij meaning black and zem meaning soil. With the soil obtaining its name from the black A horizon, rich in decaying remnants of steppe vegetation, its name in the newly developing soil science became very popular to denote a fertile steppe soil. After seven years of intensive field and laboratory work, Dokuchaev published the book *Russian Chernozem* (1883). Here, we emphasize that he had a lot of help from a team of cooperators under his supervision. Together they described the parent material and natural vegetation when the site was not a field and covered with more or less untouched natural vegetation. The slope of the terrain was also measured, and meteorological data were taken either from a nearby existing station or from measurements started just at the location where the access soil probe was dug out. Samples were taken from the horizons and analyzed physically and chemically at the contemporary levels of accuracy and technical equipment. It was the first time that an entire soil profile down to the depth of parent material was studied in the field and the observed results accompanied by those obtained in the laboratory. The final evaluation of the soil was considered as the result of parallel acting factors: parent rock, vegetation, geomorphology, and climate. Since soil evolution depends upon the length of action of these four factors, time was also included as a soil-forming factor. If one or more of the first mentioned four factors change, soil is also gradually changing. The book was translated into

French 6 years later. Since Dokuchaev recognized soils as natural bodies in evolution, he played a similar role parallel to Darwin's discovery of the evolution of living organisms, and it is universally accepted to call him the Darwin of soils.

It was not just by a chance that Dokuchaev formulated soil as a system in evolution and as the product of actions of several dynamic factors. We should not forget that Dokuchaev started his professional career as a PhD student in mathematics and physics where objectives and formulations belong to procedures for verifying objective analyses of specific systems. When he continued as a university employee of mineralogy and Pleistocene research, he followed those rational principles. Consequently, when he began investigating alluvial soils in the steppe region of Russia, he devised exact definitions and realistic sequential components for the basis of their evolutionary development within the alluvial landscape.

Today it seems reasonable for us to expect that any new body appearing on the boundary between lithosphere, hydrosphere, atmosphere, and biosphere should carry the signs of influence of each of the spheres. But we should remember that our view has been formed by generations of soil scientists who aimed their studies at the recognition of factors clearly defined by Dokuchaev.

Dokuchaev's fate did not belong every time to lucky endings. He was appointed by the administration to organize a soil science expedition to the steppe regions of Nizhniy Novgorod and Poltava. The aim of the expedition was to classify the various soils in the region and develop criteria for taxation in accordance with soil qualities. Since it was the time of a great famine, his task was to answer two additional questions: why were the yields sinking and why did the famine start? As he crossed the terrain working diligently within each and every local landscape, he planned and founded unexcelled field experimental research stations without global parallel. But when Dokuchaev had all field stations ready to collect the data and to profit from the evaluated observation of the small soil hydrologic cycle, the long-lasting period of famines ended and the top bureaucrats behaved in the same way as they are behaving nowadays everywhere. Low yields ended, taxes were again readily paid, there was no longer a loss of anticipated state income, poor workers as well as poor farmers in the countryside no longer had their homes burned down owing to long-lasting extreme weather conditions, and no luxurious houses of the rich were in danger of flames – the countryside was again quiet.

With the top bureaucrats not wanting to spend any more of their discretionary money supporting research of soils even for the benefit of farmers and others in the countryside, all money supporting the field research was locked. "No money available," declared the government. Coming just when the first results of his inventive, truly comprehensive field study were to be expected, Dokuchaev was deeply disappointed by the declaration. In spite of it, he wrote in 1892 the book, *Russian Steppes: Study of Soils in Russia, Their History and Presence.* And he did not abandon the idea of soil evolution. However, thereafter he sank into depression and alcoholism due to the lack of financial support of his promising field research, and he died in 1903. His importance for our understanding of soil evolution as well as soil properties and processes during the evolution is comparable to the importance of Darwin and his theory of evolution for living organisms. The formulation of evolutionary theory for inanimate objects was comparably just as difficult as Darwin's evolutionary theory for living organisms.

Dokuchaev's ideas reached world soil scientists mainly through translations. The first time that his ideas began to penetrate into the world was when they were meritoriously described in the book *die Typen der Bodenbildung* (The Types of Soil Formation), written in German by his student, K. D. Glinka, and published in Berlin, 1914. The book, translated into English by C. F. Marbut in 1927, provided another path for Dokuchaev ideas to become known outside of Russia. They are alive up to now in recent soil classification systems. But Glinka, another great scientist, ended even worse than Dokuchaev.

Chapter 13
Dear Soil, What Is Your Name?

The soil classification systems and the names of classification units (soil classes, soil types, soil orders, soil families) have been changed many times since the times of Dokuchaev and Glinka, but in spite of all of them, the scientific base remained still the same: soil names were determined according to the nature and sequence of horizons and with consideration of soil properties in individual horizons. The names, especially in Europe, were frequently adopted from popular, everyday expressions of farmers. To a certain extent many other names were partially or fully derived from classical languages keeping the main characteristics of soil evolutionary processes. Just a few examples follow.

Soils originating on limestone, marl, or dolomite were not deep and easily recognized from the presence of small pieces of sharp-edged fractured rock known as skeletons. When these soils were plowed, the skeleton particles rubbing against the plowshare caused a characteristic creaking. Because the popular word for such a noise in Ukraine and in Poland is *rendzich*, the soil was denoted as *Rendzina*. Although it could be commonly described as a creak soil, such a rough translation was not adopted.

In warm humid tropics the strong chemical weathering of silicate minerals releases oxides and even ions of Fe and Al that subsequently precipitate on surfaces of the remaining quartz and silicate minerals and causes a rusty red color to appear. With the precipitate covering mineral surfaces as well as filling small pores, the soil has an appearance similar to that of a well-burned brick. Actually, bricks were cut and hardened on sun. The Latin for brick is *later* and from it the name of soils was derived, earlier *Laterite*, now *Latosol*.

Next to the southern part of the Chernozem zone in Russia, there are soils with an intensive red-brown horizon below their A horizon. With their colors exemplifying those of chestnuts, farmers called them *Kastanozem* soils, or with a simple translation – chestnut soils. The Russian term was accepted by European soil scientists and finally by all soil scientists. Thus, we have up to now the soil type Kastanozem without translation.

© Springer Science+Business Media Dordrecht 2015
M. Kutílek, D.R. Nielsen, *Soil: The Skin of the Planet Earth*,
DOI 10.1007/978-94-017-9789-4_13

It was similar in the north cool regions, where the ash-like whitish horizon below a poor A horizon was named podzolic, since in Russian, *zola* means ash and *poda* or *pochva* means soil. The whole soil was then named *Podzol*. In a similar way the *Gleysols* or *Gleys* were named since the soil in the horizon below the shallow permanent groundwater level was *gley*-like, in translation from Ukrainian it was glue-like. Or simple sequence of A/C horizons on sandy parent rock gave rise to *Arenosol*, since Spanish *arena* is sand when translated.

Many national taxonomic systems keep some of the local names for soil properties that are transposed into the name of the soil. But in general, they have some links to the World Reference Base – a system elaborated by FAO, UNESCO, and the International Union of Soil Sciences. Owing to their adaptation to local regional situations, national systems provide more details using special terms for specific situations within local environments. The USDA Taxonomic System, or briefly US Taxonomy, is based on recent principles of taxonomy with strictly defined characteristics of soil genesis and soil profile descriptions. If the US Taxonomic System is nationally used, soil scientists have a free hand to combine the taxonomically well-elaborated units with low-level specific features many times not yet described, but open to use by the manifold taxonomic system.

Even if both taxonomic systems are painstakingly elaborated in detail, the final classification depends upon the experience of the soil scientist. He is in a similar situation as the physician, who examines the patient and obtains the results of blood analysis and other important tests. The doctor's decision about the type of illness and about the treatment depends not only upon the results of tests but also upon his experience. The soil scientist specialized in pedology also considers the results of chemical analyses and soil physical characteristics obtained mainly in the laboratory, but the examination of a soil profile observed within a pit dug into the landscape leads to a decisive identification of a particular soil taxon. Hence, two or more soil scientists may individually assign different taxonomic names to the soil that they observe in the same pit. Their decision about soil's name depends upon their practical experience and upon the soil classification system they are using. We shall describe two internationally dominant systems. The national soil classifications go into more detail, and they are usually based on principles of one of the two international systems we describe in the next two chapters. Be they physicians diagnosing illnesses or soil scientists identifying soil taxons, their success treating patients or managing soils is somewhat probabilistic and, indeed, never completely guaranteed for short nor long times into the future.

13.1   World Reference Base (WRB)

WRB is formulated using descriptions of soil properties within individual horizons composing the soil profile. Our selection of those diagnostic horizons contributes to our objective understanding of soil-forming processes. The prototype for our conscientious selections can be traced to the pioneering influence of Dokuchaev's approach that now defines the highest level or category in the system. A total of 32 reference soil groups, known as RSGs, are now formulated. The first edition of WRB (1998) was comprised of 30 RSGs. Today, the number has been extended to 32. The nomenclature retains terms used traditionally and usually related to current languages. The lower-level units are expressed by exactly defined prefix and suffix qualifiers that are related to secondary soil-forming processes.

The RSGs are sequenced into ten sets according to dominant identifiers. We now elucidate each set.

1. Organic soils with a dominant presence of organic materials at various degrees of humification. The influence of abundant organic materials accounts for specific soil features that separate these soils from mineral soils.

Histosols are the one and only RSG belonging to this first set. Its name was derived from the Greek *histos* meaning tissue. These soils have more than about 15 % of organic carbon in the top 40 cm even though, according to WRB, the top horizon may be of much greater thickness. According to the US Taxonomy, the content of organic matter is more than 25 % and typically increases with time. Due to a high content of semi-decomposed organic materials, the bulk density is very low. Let us remind ourselves that wood floats in water only because of its low density. The soil is poorly drained since organic matter holds water very well and the products of humification contribute to high values of contact angle and hydrophobicity. Hence, the wettability of a Histosol is low. Once it becomes dry, it accepts water very slowly and wets with difficulty. For example, the rain infiltration rate after a dry summer is practically zero with raindrops easily rolling across its surface causing a high runoff without measurably wetting the surface mixture of soil organic matter and the more or less decomposed organic remnants. Very low hydraulic conductivity values further contribute to waterlogging of the once water-saturated Histosol and subsequent anaerobic processes together with prevailing acidic reaction. If they are used in agronomy, which is not very frequent, they are usually managed as a pasture. The majority of Histosols occur in boreal and subarctic regions and in lowlands of mountainous areas. If Histosols are drained and brought into agricultural production, they should be regularly limed in order to reduce the acidity and in order to bring various plant nutrients into chemical forms more acceptable by cultivated plants. But generally, it is more advantageous to introduce plantation cropping and eventually forestry instead of annual cropping. Histosols are known in some countries in their national classification systems as, e.g., Organosols, Moore (Niedermoor, Hochmoor), Peat Soils, Muck Soils, or Felshumusböden. The US Taxonomy retains the identification of Histosols as the name of a soil order term.

2. Since WRB considers human activity as a soil-forming factor, it separates a special set of soils where human activity dominates over other factors, some of which do not even have a chance to play a peripheral subordinate role. The system recognizes two RSGs in this set: *Anthrosols* and *Technosols*.

Anthrosols are soils which have been profoundly modified by human activities, and their name was derived from the Greek *anthropos* meaning human being.

The dominant human influence comprises the deposition of organic wastes and irrigation and cultivation of loose powderlike to fine granular material that would not be denoted as soil under natural conditions. As a result of continued manuring over long periods of time, the surface horizon has features which may reach to depths greater than that of the humus A horizon in soils evolved under the natural conditions of that region. With evidence of old cultivation and modification of the relief by farmers, the physical properties of these soils are usually favorable in spite of their earlier less favorable chemical characteristics. Today, systematic fertilization and careful management of soil pH sustain the fertile character of the majority of Anthrosols except those in the semiarid zone with a high salinization hazard. They include soils of national classification systems: Plaggen soils, Paddy soils, Oasis soils, Terra Preta do Indio, Agrozems, Terrestrische anthropogene Böden, and Anthroposols.

Technosols denote soils that originated by technical skill of humans; the term has its roots in the Greek *technikos* meaning skillfully made. The parent material is selected in accordance to the aim decided by man. There is no profile development except for some humus-containing soil occasionally deposited on the surface as a top layer covering the dumped material. This top layer enables the revegetation of materials filling in surface mines or covering refuse dumps, oil spills, coal fly ash deposits, and similar by-products of technological processes. Technosols were originally not recognized as a RSG. But the permanent increase of their area and their extensive influence upon the environment required their special introduction as a separate RSG.

3. The third set is formed by soils with a severe limitation to rooting. There are two RSGs in this set: *Cryosols* and *Leptosols*.

Cryosols are mineral soils formed in permafrost where water, if present, occurs as ice, like small ice lenses or as small honeycomb structures. The name was derived from the Greek *kryos* meaning cold. Parent rock was formed by a variety of loose unconsolidated materials like glacial and colluvium sediments deposited at the base of slopes or by wind and river sediments deposited in earlier geological epochs. The thickness of parent material is highly variable and is associated with a sparse tundra vegetation and lichen coniferous or mixed forests. The soil profile is composed of an active surface horizon that thaws every summer and protects the underlying permafrost. Human activities like oil and gas mining or introduction of agriculture, transport constructions, as well as natural changes lead to material flow and mixing (cryoturbation) accompanied with erosion of thawed soil, volumetric change, thermal cracking, and great surface morphological changes. The national classifications use the terms Cryozems, Cryomorphic Soils, or Polar Desert Soils. They are as Gelisols in US Taxonomy Orders.

Leptosols are very shallow gravelly soils overlaying a continuous rock or gravel layer with less than 10 % of fine weathered material or with continuous hard rock within 25 cm from the soil surface. Provided that their stony character persists, they exist in all climatic zones including permafrost. They are widespread in mountains. The name has its roots in the Greek *leptos* meaning thin. Due to their abundant gravel content, they have high hydraulic conductivity values and manifest rapid internal drainage. As a result, they are sensitive to drought during rainless periods even in humid climate. Although they are suitable for grazing, they are much more beneficial to forestry. The national systems recognize them as Petrozems and Litozems or as subtypes Leptic Rudosols and Tenosols. If they are on calcareous rocks, they belong to subtypes of Rendzinas that earlier belonged to the principal soil types owing primarily to their neutral or slightly alkaline reaction (pH) thus forming a positive medium on which to plant the great majority of agricultural plants. Even now some national systems keep them on the top classification level using a variation on the original term like Eurendzina. The US Taxonomy recognizes Leptosols as Lithic Entisols at a lower taxonomic level. Rendzina soils belong therefore within the order Entisols to great groups.

4. Soils strongly influenced by water. The influence varies in many different aspects. There are alternating wet and dry conditions that highlight the important role of swelling and shrinkage. There are regularly flooded soils. There are soils influenced by fluctuating high groundwater level. There are soils with high concentrations of dissolved salts transported by the flux of soil water. As a result, we recognize the following RSGs: *Vertisols*, *Fluvisols*, *Gleysols*, *Solonchak*, and *Solonetz*.

Vertisols denote heavy clay soils containing a high portion of swelling clay minerals of smectite type whose properties also depend upon the value of exchangeable sodium percentage (ESP). During regularly occurring long-time rainless seasons, they dry out, and as they shrink, deep cracks appear and are subsequently filled by dust from neighboring soil surfaces. With the onset of rain, each individual area of heavy clay surrounded by cracks tends to swell, but the original cracks now filled by "foreign" soil material do not swell. With each small area of clay continuing to swell upward to a shape akin to a loaf of bread, the microrelief across the entire soil surface forms a shape of an ensemble of many loaves within the originally formed cracks. Such a microrelief is called gilgai. All together there exists a constant internal soil turnover, which gave the RSG name derived from the Latin *vertere* which means to turn. Alternate shrinking and swelling causes self-mulching, where the soil material consistently mixes itself, causing Vertisols to have an extremely deep A horizon. Vertisols are most frequent in semiarid tropics with average annual rainfall above 500 mm. They have high agricultural potential provided that they are irrigated and that the planted crop roots are not damaged by volumetric changes. Since they occur on large plains, they are suitable for large-scale mechanization and irrigation systems. However, tillage is hindered by the stickiness in wet soils and by hardness and large clods in dry soils. Infiltration into a dry soil with well-developed fissures is initially very high when water flows easily and quickly into large cracks, but after the cracks are filled, the infiltration through the clay surface is very slow. Owing to their soil hydrologic restrictions, the range of readily available water between wilting point and field capacity is small. Some of the local names which penetrated into several national classification systems are Black Cotton Soils, Black Turf Soils, Regur, Margalites, and Vlei Soils or in modified forms like Vertisols and Vertissolos. The US Taxonomy recognizes the order Vertisols.

Fluvisols are young soils developed on alluvial river sediments. Their name has roots in the Latin *fluvius* which means river, but they also occur on lake and sea sediments. Many of them are flooded periodically up to the recent time, and in such cases their fertility is usually reestablished. Their profile frequently manifests signs of stratification, especially in old alluvia where conditions for sedimentation were historically changing. With the accumulation of humus originating from plant remnants remaining in place, the top A horizon is more or less distinct. In the bottom part of the profile, signs of temporal chemical reduction combined with periods of aeration could be found if there is a correspondence between the water level in the riverbed and the groundwater level. In such cases, the existence of a gley horizon, easily recognized by its greenish to blue patches and low pH value, is attributed to the acidification caused by waterlogging within such a deep layer. Fluvisols generally belong to fertile soils. This favorable condition for plant production provides a logical explanation of why archeologists found the first steps of agricultural revolution in regions distinguished by Fluvisols. Only tidal soils are the exception in Fluvisols' fertility since they are usually very saline. Fluvisols appear in national classification schemes under the name Alluvial Soils, Auenböden, Sols Minéraux Brut, or Sols Peu Évolué. The US Taxonomy classifies them as suborder Fluvents in Entisols.

Gleysols have been developed on wetlands with high elevation of groundwater level that can reach up to the root zone. Occasionally, they evolved from Luvisols when the groundwater level rose during most of the year either due to changed natural conditions or sometimes due to human activity. The bottom part of the profile is fully saturated by the groundwater during the whole year. Due to the weathering of minerals and the transformation of organic matter in the absence of oxygen and under permanent anaerobic conditions, this horizon has a greenish-blue color with a gray hue. Above this G horizon is the mottled gleyish horizon where the anaerobic waterlogged conditions are regularly interchanged with aerobic conditions when water is drained from a portion of big pores and reddish or orange mottles indicate localized reoxidation of ferrous compounds in the soil matrix. These small, irregular patches are often associated with root channels, animal burrows, or cracks of the soil material when waterlogging occurs at greater depths or is even absent. On top is the humus horizon, which may sometimes be interchanged with peat or swamp. We have already mentioned that the folk word for glue in Ukraine is *gley* that was transferred to soils being overwet and sticky. If Gleysols are regularly tilled and cultivated when they are too wet, their soil structure is destroyed with their physical conditions no longer being favorable for cultivated plants. Permanent grasses or swamp forests then become preferred alternatives. When Gleysols are drained and the groundwater table lowered, the majority of soils start to belong to fertile soils, provided that the pH is kept at optimal values. Usually liming is recommended. In Southeast Asia rice is frequently grown on Gleysols. They appear in national taxonomies also under the name Gleyzems, Meadow Soils, Hydromorphic Soils, or simply Gleys. The US Taxonomy ranks them as suborders of Inceptisols (Aquepts), Entisols (Aquents), and Mollisols (Aquolls).

Solonchaks are poorly drained soils having low hydraulic conductivity and high concentrations of soluble salts within their profiles. Their existence is restricted to arid and semiarid climatic zones, and they appear only exceptionally in coastal areas. The high concentration of salts is reflected by their name from the Russian *sol* which means salt. The parent rocks of these soils are unconsolidated materials. Since the potential evaporation is substantially higher than the average annual precipitation, the dominant direction of water flow is upward and the flowing water simultaneously carries dissolved salts – the products of weathering. As the water evaporates, those salts remain in the topsoil horizons, some of which eventually precipitate as a whitish powder just on the surface. The salinization process is conditioned by the depth of the groundwater table. When the water table is below the bottom of a soil profile, salinization is relatively slow. On the other hand, salinization is accelerated when the water table rises into a soil profile that eventually develops distinctly visible gleyic marks within its horizons. Greater accumulations of salt occur in low-lying topographic areas owing to their transport from shallow water tables. If soils of arid and semiarid regions are irrigated, salinization is a potential hazard even in the absence of a groundwater table. Carefully performed sprinkler and drip irrigation methods have the potential to restrict or completely avoid salinization and Solonchak development. Using irrigation water containing low amounts of salts is also an important prevention of salinization. Certain plants evolved in arid and semiarid zones, like cotton, resist soil salinity, and even thrive up to a threshold value. But for even these plants, their yields decrease whenever the soil salt content rises permanently above this threshold value. Solonchaks are excluded from agricultural use after their salt contents cross a critical value defined for a group of selected plants in a specific, local environment. A traditional weapon against salinization is flush irrigation – the repetitive application of excessive amounts of water to create a vertical downward flux during a selected time period to adequately flush deleterious salts out of the soil profile. The weapon has to be used judiciously, i.e., it has to be accompanied with intensive drainage, allow an aftermath of the groundwater table being at a desired critical depth below the root zone, and connect with a successful method of transporting the salty drain water to another acceptable location at or below the surface of the landscape. In old national systems we could find Alkaliböden or Weissalkaliböden, Saline Soils, White Alkali Soils, or Salt-Affected Soils. They belong to Orthids in the order of Aridisols and to several taxons of low taxonomic level of Entisols in the US Taxonomy.

Solonetz is soil with a high content of exchangeable Na^+ and Mg^{2+} ions that create an alkaline pH reaction of about 8.5. The name again derived from the Russian *sol* which means salt and coupled with *nec* which means expression of a negative property obviously refers to unfavorable soil physical characteristics for cultural plants, e.g., destruction of soil structure, strong swelling and shrinkage, and an extremely low hydraulic conductivity within the horizon having a high percentage of exchangeable Na^+ and Mg^{2+}. The concentration of soluble salts in top horizons is substantially lower than that in Solonchak because they were washed out. This condition could be an alternative nuance of the Russian *nec* indicating that there are no more soluble salts of high concentration, i.e., a negative of salts. Na^+ and Mg^{2+} were left as traces of the earlier accumulation of salts. The permanent "surviving" of Na^+ and Mg^2 was enabled by their fixation on exchangeable positions of the solid particles, mainly smectites. Solonetz have evolved on unconsolidated materials of flat lands having a climate typified by long hot rainless summers. Annual precipitations are higher than those linked with Solonchak, but still below 500 mm. If Solonetz evolve from Solonchak during long-term irrigation, then the main factor influencing the process is high content of sodium salts either in earlier Solonchak parent material or due to the use of Na^+-rich irrigation water. The soil profile reflects the dominant process of washing out the soluble salts. Due to the decreased salt concentration and prevalence of exchangeable Na^+ (partly also Mg^{2+}), the clay particles cannot coagulate and stick together. Inasmuch as they exist in a peptized state of individual colloidal units having diameters well below 1 μm, they are simply transported out of the top horizons through slightly bigger sized pores. This eluviation of tiny colloids is the dominant mechanism for a thin natric horizon to develop below a black or gray brown surface humus horizon. Its gray whitish color, its high pH value (about 8.5 or even more), and its columnar aggregates with rounded tops are typical properties of natric horizons. Signs of gleization could appear at the bottom of natric horizons. When wet, they are dispersed and sticky and have very low hydraulic conductivity and low air permeability. When they are dried, they form a hard crust inside their soil profile and wide cracks on their surface. Solonetz could be brought into agricultural use only after an expensive amelioration that starts with planting special Na-resistant grass and incorporating gypsum or calcium chloride in the topsoil. Next, gradual deepening soil tillage mixes the ameliorated topsoil with the upper part of the natric horizon. As Ca^{2+} cations replace Na^+ in their exchange positions, some coagulation begins and the first step of aggregation takes place. The amelioration is expensive and requires an extremely large number of years to obtain only partially acceptable results. Hence, it is not surprising that the vast majority of Solonetz have not been reclaimed – they are either used for extensive grazing or left lying fallow. The old classifications denoted them as Alkali Soils, Sols Sodiques, or Schwarzalkaliböden. In the US Taxonomy, Solonetz corresponds to sodium-rich Aridisols and Mollisols.

5. The fifth set of soil groups is the RSGs in which iron (Fe) and/or aluminum (Al) chemistry plays a major role in their formation: *Andosols*, *Podzols*, *Plinthosols*, *Nitisols*, and *Ferralsols*.

Andosols developed mainly on volcanic ashes and glasses but also on tuff, pumice, cinders, and other volcanic ejecta in the circum-Pacific zone (New Zealand, the Philippines, Japan, Ecuador, Peru, etc.). They also occur in humid climate on other silicate-rich materials. Rapid weathering of porous materials results in stable complexes of minerals with humus. The top A horizon is dark to black, with fine structure. The color gave the name to those soils, stemming from Japanese *an* meaning black and *do* meaning soil. The clay fraction is composed mainly of allophane, the major weathering product of volcanic materials with an admixture of halloysite. This composition predetermines the properties of the A horizon. The soil is very porous mainly due to the spherical shape of allophane. It is well aggregated with many fine roots and has a nearly neutral reaction and a very low value of bulk density. Its profile gradually changes color within the transition to parent volcanic material. This transition, sometimes not properly denoted as B horizon, is gray brown and maintains high porosity and a high hydraulic conductivity. However, if there is a horizon with more intensive weathering below the A horizon, it is appropriately designated as B horizon. Below it at the bottom of the profile, the parent materials appear (C horizon). The profile of Andosols is more complicated if there is a repeated deposition of fresh ash that is forming different horizons. Andosols are fertile and easy to cultivate and have good water storage and high infiltration. Moreover, they are easily penetrated by roots. Their single problem may be phosphate fixation caused in certain instances by active Al and Fe. This unwanted property is reduced by applications of lime and organic material. A great variety of crops are planted on Andosols. Paddy rice cultivation is a major land use in lowlands having a shallow water table. Some national classifications recognize them as Black Dust Soils, Vitrisols, or Volcanic Ash Soils. In the US Taxonomy, Andosols are known as Andisols.

Podzols are soils with an ash-light-gray horizon below a top humus horizon. The name was already explained at the end of the introduction in this chapter, and if roughly translated from the Russian, it would be ash-like soil. Since the original vegetation was heather and coniferous forests, the organic material for humification already carried an acid reaction. With humid and cool climatic conditions both contributing to existing acidification processes, the acid reaction caused intensive weathering with free release of Fe and Al. The soluble metal-humus complexes commonly known as chelates were formed as humification produced an abundance of fulvic acids in the presence of mineral particles. The prevailing downward rainwater percolation caused intensive leaching and the evolution of an ash-like eluviated horizon just below the humus horizon. Carboxylic and phenolic groups of low molecular weight humic and fulvic acids act as pliers grabbing the Fe and Al until they become "saturated" complexes. Being soluble, they are washed down until the organic part of the molecule has no more "thirst" for ions, at which time the complex material is no longer stable as it was during its transport. The eluviated material is therefore deposited in the next illuvial horizon typically characterized by dark brown or rusty brown hues. Dark gray up to black-gray streaks could appear with accumulation of low molecular humus. In very humid tropical climates, Podzols can also develop in a different way with their genesis starting and bounded by a permanently washing and flushing out soil water regime. The parent rock of sandstones and their weathering products are poor in clay minerals and plant nutrients. They are washed downward into the profile usually with many trace elements already being captured into the transported metal-humus complexes. All Podzols have an acid reaction of pH well below 6 and even down below 5 and hence manifest Al toxicity and are also deficient in P. Their amelioration must start with neutralization of the acid reaction, deepening of their plowed topsoil, and drainage and removal of excess water. Then and only then could fertilization begin. The great majority of national systems use the name Podzol. The US Taxonomy refers to them as Spodosols.

Plinthosols are mainly tropical soils exemplified by an iron-rich, in some cases also manganese-rich, humus-poor mixture of kaolinitic clay and gibbsites. The consequence of this composition is a very small cation exchange capacity. When this material exists in various forms from soft lumps to hardpan, it is described as plinthite, and when it extends to continuous rocklike forms, it is denoted as petroplinthite. Laymen could confuse them with ironstone. The name was derived from the Greek *plinthos* meaning brick. They are also known as Lateritic soils or Latosols where the name was derived from the Latin *later* meaning brick. They developed during extremely severe weathering conditions in wet hot tropical climates. As weathering proceeded, bases were released and removed, while sesquioxides accumulated within the profile at the place of weathering. The process intensified when Fe in its bivalent soluble form was transported by groundwater from neighboring slopes to a lower flat terrain or by a great downward fluctuation of groundwater table, causing a precipitation of Fe_2O_3 and giving an intensive red color to the Plinthosols. They are more frequently and more vividly developed from basic rocks than from acidic rocks. We recognize them in the field by red mottles that are firm when they are wet and very hard when dry. In either case, they are difficult to cut with a knife and do not stain our fingers when we rub them. In some instances the Plinthosols originate on old, many meters thick petroplinthites. The profile of Plinthosols is characterized by the presence of illuvial B horizon and even with the eluvial E horizon above B horizon. Their sequences of horizons are designated as A-B-C or A-E-B-C, respectively. Inasmuch as Plinthosols are shallow soils, their agronomic use is primarily constrained by factors affecting plant nutrition. Their disposition to retain adequate amounts of macronutrients is constrained by their extremely small cation exchange capacity (CEC). The existence of plinthite restricts the rooting depth of plants. Stoniness is frequent. With very low contents of both the macro- and micronutrients, their natural fertility is usually poor. These soils are known in some national systems as Groundwater Laterite Soils, Lateritas Hydromorficas, or Sols Gris Latéritiques. The US Taxonomy recognizes them as Oxisols and specifies them at lower taxonomic levels.

Nitisols are deep well-drained reddish and red-brown soils of tropical rain forest and high savanna with a typical nitic horizon below the relatively fertile and humus-rich top horizon. The name of the RSGs is related to nitic horizon with its name derived from the Latin *nitidus* meaning shiny, since the stable polyhedric aggregates with strong angular structural elements have many very shiny walls. Since mottling of the nitic horizon is always lacking, there is no influence of periodic waterlogging on the development of these soils. This absence of mottling is one of the signs of how Nitisols differ from Plinthosols. However, there is still a dominance of kaolinite and halloysite in their clay fraction and their cation exchange capacity (CEC) is still very low. Although the dominant color of Nitisols is red, it is less intensive than that of Plinthosols. Nitisols, composed of A-B-C horizons, are profitably used for agricultural crop production. Even though their level of plant-available P is low owing to a high content of free Fe, they are rich in soil microfauna and well homogenized without sharp boundaries between horizons, and all of their biologic diversity contributes to good physical properties. They are recognized in national systems also as Terra Roxa Estruturada, Sols Fersialitiques, Ferrisols, or Red Earths. In the US Taxonomy they belong to Ultisols and to Alfisols, e.g., Kandiudalfs.

Ferralsols are deep red or yellow soils of humid tropics that differ from Nitisols by the absence of a nitic horizon. We can judge from their name that they are rich in sesquioxides (Fe_2O_3, Al_2O_3) and perfectly crystallized kaolinites are predominant in their clay fraction. Many local names refer to their color. The name has roots in the Latin *ferrum* and *aluminium*. The red color is due to thin films of hematite covering the sand and silt particles and sooner or later fills parts of ultra-micropores. If a yellow color occurs, it is due to the dominance of goethite, a mineral named in honor of a great writer and scientist. Although he did not discover the mineral, the famous German poet was not only a great personality in biology and mineralogy, but he also entered the discipline of soil science and other complicated geosciences through the name, goethite. Goethite is found in every soil type where increased concentrations of iron oxides appear in a horizon. It is transformed into limonite after hydration or into hematite when dehydrated. The terrain of Ferralsols is level or moderately undulating of Pleistocene age or more, i.e., 2 million years or more. The weathered parent material of the same age has a stable microstructure. Due to the stable mineralogy and fixing properties of sesquioxides, the microstructure of Ferralsols is also stable. Even though their physical properties are good, owing to a very limited stock of plant nutrients together with an unfavorable fixing of phosphates similar to Nitisols, their natural fertility is low. Moreover, owing to their acid reaction, they are plagued by Al toxicity. In national classification systems they are named, e.g., as Latosols, Lateritic Soils, Sols Ferralitiques, or Ferralitic Soils. They are denoted as suborders and great groups within the order of Oxisols in the US Soil Taxonomy.

6. Set of soils with perched water include two RSGs: *Planosols* and *Stagnosols*. In both groups, waterlogging plays a principal role in their development.

Planosols occur in plains. They have a coarse-textured surface horizon with a sharp transition into alluvial deposits typical with all signs of permanent or regularly seasonal waterlogging. The name was derived from the Latin *planus* meaning flat. Their profile reflects geological stratification more than soil evolution. Owing solely to stratification, they typically have a very low permeability. They are used for planting rice as a single crop in Southeast Asia, but they require special care. In other regions, even if they are drained, they offer usually poor yields. Their negative features for agronomic use are less extreme if the waterlogging is only seasonal. They appear as Pseudogley Soils in several national classifications. They are equivalent to Albaqualfs, Albaquults, or Argiabolls in the US Taxonomy.

Stagnosols are soils periodically waterlogged due to surface water in regions with humid to perhumid climate in flat or gently sloping land on various sedimentary materials with very low hydraulic conductivity. The name was derived from the Latin *stagnare* meaning to flood. They usually have mottles from the topsoil to the subsoil. The color patches are from greenish to dirty blue, irregularly interrupted by bleaching which appears also in the top A (humus) horizon. All of their identifying characteristics are due to stagnating water during wet rainy periods causing chemical reduction. Stagnosols differ from Planosols in that the long-lasting water saturation is not caused by an abrupt textural change in the profile. Their agricultural use is rather limited since a simple pipe drainage system is not sufficient to improve their lack of aeration and oxygen deficiency. In other words, owing to their hydraulic properties, a substantial increase of air-filled porosity, mainly of medium and coarse pores, is simply not attainable. Many national classifications denote them as Pseudogleys. The US Taxonomy incorporates them in Aqualfs, Aquents, Aquolls, in taxons starting with the Latin *aqua* meaning water and in Inceptisols.

7. Soils evolving in steppe regions with humus-rich top horizon. Rich in base saturation, their vertical transition to parent rock is gradual: *Chernozems*, *Kastanozems*, and *Phaeozems*.

Chernozems are probably the most popular soils and are ranked with the best of them owing to their high natural fertility. The term was introduced by the founder of the systematic studies of soil types, the Russian V.V. Dokuchaev. The soil profile is simple – a humus-rich A horizon having a thickness ranging from 30 to 100 cm covers the usually silty loam to loamy loess parent material. The high content of humus reaching to about 10 % is readily apparent from the black color. Their name was derived from the Russian *chernij* (*chorny*) meaning black and *zemlia* meaning earth or soil. The high content of humus is the result of two positive factors: the dense steppe vegetation consisting mainly of grasses, especially papilionaceus plants, and mild climate with cool winters and long warm summers with low frequency of rains. Since the climate forms a continuous belt and the same is valid for the zone of steppes, earlier classifications recognized Chernozems as zonal soils. Their richness in high-quality humus causes stable aggregated soil structure and favorable soil physical conditions for plants. These attributes with those of calcareous loessial parent material provide a neutral chemical reaction needed for an ample supply of readily available plant nutrients. Hence, Chernozems are universally known for their high fertility and noteworthy crop productivity. The term Chernozem was taken over in majority of national classifications, sometimes modified like Chernosols in Canada, or Chernossolos, and in some instances in literal translation, like Schwartzerde and Black Earth. In the US Taxonomy we find Mollisols and their suborders.

Kastanozems are soils in many aspects similar to Chernozems, and even their classification as zonal soils was the same. Their zone, to south of the Chernozem zone in the Northern Hemisphere, differs from the environment of Chernozems by substantially drier climate and shorter grass steppe vegetation. With the majority of their plants requiring less soil water, they are basically more resistant to dry climatic periods. Kastanozems show more intensive accumulation of secondary carbonates, since the net upward unsaturated flow prevails slightly over the net downward flow of rainwater. Consequently, they have a dark brown surface horizon, which is less deep and less black than that of Chernozems. They also differ from Chernozems by having an additional horizon in their profile – a B horizon with a cinnamon to a more pale color. Sometimes there is an accumulation of lime or gypsum. Their name results from dark chestnut color in A horizon, since the Latin *castanea* or the Russian *kashtan* means chestnut and *zemlia* means earth or soil (see also Chernozem).

When compared to Chernozems, they dry out to greater depth in dry summer seasons and are not as completely wetted by rains. Their structural stability, mainly that of microaggregates, is weaker, and their hydraulic conductivity is less favorable. Sufficiently high yields are reached only if they are irrigated. Their synonyms in national systems are either translations or names that are derived from colors such as Chestnut Soils, Kalktchernozems, or Brown Soils. They belong to various taxonomic levels of Mollisols (Borolls, Ustolls, Xerolls) in the US Taxonomy.

Phaeozems are soil in the zone north of the Chernozem zone in the Northern Hemisphere. They were formed in a wet steppe or prairie zone. Being more intensively percolated and leached, their humus horizon is less rich in bases when compared with the topsoil of Chernozems and Kastanozems. Moreover, we don't find secondary carbonates in the top 1 m of their profile. The name was derived from the Greek *phaios* which means dusky. The Russian *zemlia* was explained earlier, it means earth or soil. We find in their profile the cambic or argic B horizon that is slightly more rich in clay than the A horizon and usually has a distinct polyhedral structure with aggregates having more than 6 slightly irregular walls. This B horizon is between the humus A horizon and parent material formed by loess or loessial loam and other unconsolidated basic fine-grained materials (C horizon). Phaeozems are fertile, excellent farm soils, well reacting to additional irrigation. They are usually identified with the color brown in many national classifications, like Brunizems derived from the French *brun* meaning brown and Parabraunerde derived from the German *braun* meaning brown or Fahlerde. The US Taxonomy keeps them as Udolls and Albolls in Mollisols.

8. The next set comprises soils from drier regions with accumulation of a certain material. If it is gypsum, the RSGs are called accordingly *Gypsisols*. If it is silica, the RSGs are *Durisols*. And if the accumulated material is calcium carbonate, the soils are *Calcisols*.

Gypsisols are soils existing in dry areas of semiarid regions or even in desert areas. Due to the climate with precipitations substantially lower than evapotranspiration, carbonates and sulfates are transported immediately below the thin top horizon where gypsum ($CaSO_4 \cdot 2H_2O$) accumulates and precipitates during long-lasting hot and rainless summers. This accumulation was the logic for the name of the soils having common roots with the Greek *gypsos* meaning gypsum. The parent material is formed by unconsolidated alluvial and wind deposits of base-rich material, i.e., materials with an alkaline reaction. Their natural vegetation is rather sparse and dominated by thorny shrubs. Their agricultural use is conditioned by cautiously managed irrigation to avoid the secondary salinization during long periods of crop production. The national classifications use the terms Desert Soils, Yermosols (Spanish *yermo* meaning desert), and Xerosols. Among the soils of the US Taxonomy, the most frequent are Gypsids in Aridisols.

Durisols occur mainly on old arid and semiarid surfaces with a relatively permeable material, but this positive feature is frequently disturbed by a hardpan of secondary cemented silica (SiO_2). The name was derived just to indicate this property, since the Latin *durus* means hard. The hard horizon (hardpan) disturbs the rooting of plants and prevents a desirable distribution of water within the soil profile whenever Durisols are irrigated. This RSG was only recently introduced into the WRB system. They are recognized in national classification systems also as hardpan soils and dorbank and in the US Taxonomy as Durids in Aridisols.

Calcisols are soils with a distinct accumulation of lime as a horizon in soil profiles in arid and semiarid climatic zones. On the occasion of the soil sometimes being wetted by a rarely occurring rain, dissolved lime is transported by evaporation flux. The calcic horizon may be hardened into a rocklike formation, i.e., a petrocalcic horizon. They occur at various depths from very shallow to deep levels within the profile, and owing to high contents of adsorbed Ca^{2+}, they are quite permeable. Their calcareous, alluvial, or wind-deposited parent materials are all further weathered very slowly in the dry climate. The natural vegetation is sparse – formed by shrubs and trees and also partly by ephemeral grasses. Generally, droughts lead to a slowdown or even to a virtual stop of soil-forming processes such as chemical weathering, accumulation of organic matter, and translocation of clay. If some traces of horizons are visually apparent, they could be the result of polygenetic evolution in a previous different climate. Calcisols are frequently rich in stones. Their name was derived from the Latin *calcarius* meaning calcareous. If irrigated, they are susceptible to salinization and potentially change into Solonchaks. Calcisols are classified as Desert Soils or Yermic Soils with the term derived from the Spanish *yermo* meaning desert in some national taxonomic systems. When Calcisols are attacked by accumulation of salts in areas without surface drainage, they are also classified as Takyrs – the name derived from the Kazakh or Turkish languages for flat land that is occasionally inundated.

9. Set of soils with a clay-rich subsoil where the following RSGs belong:
   Albeluvisols, *Alisols*, *Acrisols*, *Luvisols*, and *Lixisols*.

Albeluvisols are soils that have a thin dark humus A horizon lying on top of a leached subsurface horizon known as an albic horizon. The name of the horizon was derived from the Latin *albus* meaning white, while the second part of the RSG name "luvisols" has its roots in the Latin *eluere* meaning to wash out. The albic eluviated E horizon is an iron-depleted material changing in a sharp undulating manner into an underlying compact illuvial clay B horizon. The harsh climate of cold winters, short and relatively cool summers, and high annual precipitations cause periodic over-wetting of soil and leaching of sesquioxides together with clay particles from the eluviated E horizon into the illuviated B horizon with features typical of periodic waterlogging. The high clay content of the B horizon contributes to the origin of those features. These soils are acidic with low nutrient content, are very shallow for adequate root penetrability, and require drainage for crop production. Hence, their agricultural utility remains problematic. Some national classifications recognize them as Podzoluvisols or Orthopodzolic Soils. The US Taxonomy classifies them under Alfisols like Glossaqualfs, Glossocryalfs, and Glossudalfs.

Alisols occur mainly in tropical and subtropical environments with high amounts of exchangeable Al^{3+} cations released due to the hydrolysis of simply weathered primary minerals rich in Al and of secondary clay minerals, like vermiculite and smectites. They are acidic with limited drainage of the subsoil containing a majority of smectites in clay. A relatively high concentration of free aluminum and manganese causes their toxicity for cultural plants. Their profile consists of a thin humus A horizon usually poor in organic matter content and with an unstable soil structure. Below it is a reddish yellow B horizon with the color caused by the illuviated material. The parent material is usually rich in active clays like smectites, but their relation to low-activity clays like kaolinite decreases in individual horizons according to weathering intensity. As it increases, the concentration of free Al increases together with accompanying negative properties. If Alisols are used for crop production, they need liming to neutralize their acidic reaction and to reduce the solubility of simple inorganic substances containing Al. Nevertheless, they are still hampered by toxic concentrations of Al and Mn. And even with logical and vigilant management practices, they remain easily eroded owing to a lack of surface coherence. They exist in national classifications as Red-Yellow Podzolic Soils, Fersialsols, and Sols Fersiallitiques Tres Lessivés. The US Taxonomy recognizes them as Ultisols.

Acrisols differ from Alisols by their high content of low-activity clay minerals like kaolinites having a low base saturation. Their reaction is also very acidic. But similar to Alisols, they have a thin light-colored surface humus A horizon overlying a whitish to yellow eluviated E horizon with weak structure – sometimes without any aggregation and strongly massive. It overlies a reddish to strongly yellow illuviated B horizon. The parent materials are acid rocks and their weathering products. Acrisols are also poor in plant nutrients and have strongly bound phosphates. Their acid reaction accounts for their Al toxicity to plants. When covered by forests, their physical properties contribute positively to regional hydrologic cycles with rain infiltration, runoff, and deep drainage entirely adequate, but after being shifted to agronomic use, their physical conditions deteriorate causing poor infiltration and periods of severe soil erosion. They associate with nationally classified soil types like Red-Yellow Podzolic Soils, Sols Ferrallitiques Fortement Désaturés, and Red and Yellow Earths. The US Taxonomy recognizes them as suborders in Alfisols and Ultisols.

Luvisols are soils of the mild, cool climatic zones north of the Chernozem zone in the Northern Hemisphere. The average annual precipitations are usually slightly higher than those of evapotranspiration. Parent materials are unconsolidated sediments like loess and loessial loams or alluvial and colluvial loams. The humus-rich top A horizon is slightly leached, and active clay minerals like smectites and illites are transported by percolating water to the intensive brown illuvial B horizon. Even if the soil textures of A and B horizons differ substantially with the B horizon more rich in clay fraction, waterlogging is absent in typical Luvisols. If present, the subtype's name reflects this fact. National and older classification systems used the terms Pseudopodzolic Soils, Parabraunerde or Braunerde (Brown Soils), Sols Lessivées, and Gray-Brown Podzolic Soils. The US Taxonomy groups them into Alfisols.

Lixisols are strongly weathered soils in which clay was washed out of the eluvial E horizon to form the illuvial B horizon. Although E horizon is usually very hard, it is sometimes not easily recognized. Their name is related to the Latin *lixivium* meaning solution, lye. The relatively rich illuvial B horizon contains low-activity clays with a moderate to high base saturation. Since kaolinites and halloysites prevail, the cation exchange capacity is low in the entire soil profile. The subtropical to tropical climates of these soils differ from the climates of the majority of other ABC soils owing to winters being dry while summers being humid. Lixisols belong to old soils that started to form earlier than the Pleistocene more than 2 million years ago during conditions much more humid than those of today. Their reddish to yellow color is the result of dehydration of ferrihydrite to hematite (rubefaction) due to the existence of prolonged dry seasons. Although Lixisols have low levels of readily available plant nutrients, owing to their high content of bases, they lack the Al toxicity manifested by Ferralsols and Acrisols. They are well drained and their water-holding capacity is better than that of Ferralsols and Acrisols. They are also classified in national taxonomies as Red-Yellow Podzolic Soils, Argissolos, Sols Ferrallitiques Faiblement Desaturés Appauvris, Red and Yellow Earths, and Latosols. They form subgroups of Alfisols in the US Taxonomy.

10. The last set is reserved for relatively young soils or soils with very little or no profile development. They are grouped together with very homogeneous sands: *Umbrisols*, *Arenosols*, *Cambisols*, and *Regosols*.

Umbrisols are soils without a distinct profile development or with a relatively fresh soil with a great accumulation of organic material only partly humified, described frequently as raw humus or as mor, moder, and mull. Their name is related to the Latin *umbra* meaning shade. This organic A horizon is not mixed with mineral portion of soil, mainly sand in the C horizon on the bottom of the soil profile. Due to the combination of acidity, low temperature, and excess surface wetness, there is a slow biological transformation of plant remnants. Since Umbrisols were never cold or wet enough to have developed a distinct humus horizon of Histosols, they are classified as a separate RSG. Some national systems describe them as Sombric Brunisols and Humic Regosols and Brown Podzolic Soils. The US Taxonomy ranks them in lower taxons in Great Groups of Entisols and Inceptisols.

Arenosols are various forms of sands that originated in place either by climatic weathering of quartz-rich sediments, rocks, and wind-deposited sands or by earlier seawater weathering of shallow sea shelves. Their name is derived from the Latin *harena* meaning sand and from the similar French *arene* or Spanish *arena*. The top humus A horizon is either underdeveloped or entirely missing. Although their existence does not depend upon the climate, properties and types of eventual agricultural use of Arenosols differ according to their phase of soil development that increases within climatic zones from arid to subhumid temperate to humid tropics. Owing to a lack of clay particles, all Arenosols existing in one of these three climatic zones have common physical characteristics of very high permeability, very low water retention (field capacity), no aggregation, very low cation exchange capacity (CEC), low fertility, and unfavorable hydrologic conditions for plants – even those that are drought resistant. Moreover, whenever thin hydrophobic films occasionally cover the sand particles, the hydrologic soil properties of infiltration rate and water-holding capacity are adversely modified. In the transition zone between typical Sahara desert soils and Sahelian soils, the unfavorable properties of Arenosols are gradually diminishing, but not disappearing. In national systems we find Arenosols as Classe Des Sols Peu Évolués, Sols Minéraux Bruts, Red and Yellow Sands, Arenic Rudosols, and Psamnozems. The US Taxonomy recognizes them either as suborders of Entisols (Psamments) or as subgroup in Alfisols.

Cambisols are mainly soils in the early state of differentiating the soil profile into horizons or soils that are in the stage of transition from AC profile into ABC profile. The B horizon shows the alteration relative to the lower C horizon in color and the removal of calcium carbonate when the C horizon is formed by loess. However, $CaCO_3$ could be found in B horizon, but in lower concentration than it is in C horizon in some instances. It keeps the neutral or slight acid reaction. Texturally their content of clay particles is higher than in sandy loam. The structure in B horizon is usually polyhedral and differs from the crumbling structure of the top A horizon. Hydrolysis of iron-containing minerals like biotites, pyroxenes, and amphiboles produces ferrous iron that is partly oxidized to goethite and hematite and partly exists as a free iron film covering the sand and silt particles. These "coats" give the distinct brown color to B horizon in well-developed Cambisols. There is no leaching of bases and sesquioxides. The name was derived from the Latin *cambiare* which means to change. They are fertile soils used from the start of the agricultural revolution in the mild zone. Many Cambisols belong to the most productive soils of the earth. The old classifications and some contemporary national systems recognize them as Brown Soils, Braunerde, Sols Bruns, and Brunizems due to the brown color of their B horizon. The US Taxonomy classifies them in Inceptisols.

Regosols are very weakly developed soils originating on unconsolidated mineral parent material. They lack distinct horizons. If we describe the thin top layer differing from the parent material, we use the term ochric horizon. Its name is derived from the Greek *ochros* which means pale. It has very low content of organic matter, the weathering is usually weak, or in some instances their products form a hard crust. Regosols have a very low capacity to retain water, are sensitive to droughts, and are easily eroded. With a low nutrient content sometimes approaching zero coupled with poor hydrologic relations, their agricultural significance remains virtually nil. Regosols are classified in national systems as Skeletal Soils, Rohböden, and Sols Peu Évolué. The US Taxonomy recognizes them as lower units (great groups) of Entisols.

13.2   USDA Soil Taxonomy

The first US soil classification system at the end of the nineteenth century was based on the underlying geology. Soil maps elaborated for agricultural administration and supposedly for the benefit of all farmers were little more than maps of geologic forms appearing on the landscape surface combined with an empirical guess regarding soil texture. A principal modification eventually occurred in the 1920s when

C. Marbut introduced the idea of soil evolution stemming from Dokuchaev's concept. Marbut gradually extended the Russian system to US environmental and soil conditions between 1921 and 1935. It is worth mentioning that among his contemporary soil scientists bearing PhD degrees, he never bothered to take the final exam for his PhD at Harvard where he obtained his Master of Arts. In addition to observing and understanding interconnecting links between various components of naturally occurring soil processes, he was also fluent in German and was among the soil scientists preparing the foundation of the International Soil Science Society (ISSS) at the Prague conference in 1922. ISSS was finally established in 1924 at the Fourth International Conference of Pedology. His contacts with the European soil scientists enabled him to learn of Glinka's book that he translated into English in 1927. With this readily available translation, it was easy for US soil scientists to extend these new approaches. Indeed until 1935, Marbut himself published several drafts of soil classification according to his knowledge about the new concepts. At that time the system's highest taxonomic level designated soils into three soil groups – zonal, intrazonal, and azonal. It was a unique, meritorious achievement of Marbut that US soil scientists started to play an important role in soil classification by applying and extending theoretical taxonomic principles that eventually led to the present-day US Taxonomic System.

The main objective of this system is to define all factors influencing the evolution of the taxon and its relation to a neighboring taxon. This is achieved by a strict hierarchical arrangement with flexibility in grouping taxons. The system recognizes six descending levels that start at the top level of *orders* and below them sequential levels of *suborders*, *great groups*, *subgroups*, *family*, and *series*. Each soil has its name composed of syllables characterizing all levels. For example, soils belonging to *order* Alfisol will have the last syllable, or more precisely the formation element *alf*. If it has the udic moisture regime, then its *suborder* name is *Udalf*. The name is derived from the Latin *udus* which means moist. If the soil is well drained, has a horizon rich in clay that does not have an abrupt upper boundary, has a relatively high base saturation, is relatively deep, and has only a weak horizon development, the soil belongs to the *great group* named *Hapludalf*. The name is derived from the Greek *haplous* meaning simple. When the soil has cracks in its top horizon and no other dominant properties of Vertisols, the soil fits into *subgroup Vertic Hapludalfs*. The above example shows that the US Taxonomy created a large number of new words. A brief, scientifically incomplete explanation of the highest taxonomic level terms, *orders*, will follow together with one example of one *suborder*, one *great group*, and finally one *subgroup* of many others. The aim here is just to show how new names are formed in each of the orders. The book *Keys to Soil Taxonomy* elaborated by Soil Survey Staff contains full information needed for appropriate designation of a field soil. There are also the rules for the identification of families like particle size classes, soil depth, soil slope classes, etc. Below, we do not review or explain the two lowest taxonomic levels – *family* and *series*.

But it may be worth noting that the series, the field mapping units, have remained unchanged in spite of all the shenanigans of innumerable changes in classification since the establishment of the soil survey under Milton Whitney. This is the genius of the American system: the soil series used and known by farmers and extensionists (say Miami Series) remains Miami Series.

Alfisols have the formation element *alf*. They occur mainly in humid and subhumid climates. An increased clay content B horizon occurs within their profile without a leached horizon being developed. These soils are either identical or very close to Luvisols, Nitisols, and Acrisols in World Reference Base (WRB) described in Sect. 13.1. The suborder Aqualfs is very similar to the WRB Stagnosols. An example of one of the many Alfisols is the suborder Ustalfs, great group Haplustalfs, and subgroup Vertic Haplustalfs.

Andisols have the formation element *and*. The name is derived from the Japanese words *an* meaning black and *do* meaning soil. They evolved on volcanic ash and other volcanic materials. They are identical to WRB Andosols. An example is the suborder Udands, great group Fulvudands, and subgroup Typic Fulvudands.

Aridisols have the formation element *id*. Their name has roots in the Latin *aridus* which means dry. They have been formed in arid climate and due to this factor they have low content of humus. The prevailing evaporation over poor precipitations is reflected frequently by salt accumulation in subsurface horizons. They are similar to soils in WRB system: Solonchaks, Solonetz, and some Arenosols. An example is the suborder Gypsids, great group Petrogypsids, and subgroup Xeric Petrogypsids.

Entisols have the formation element *ent* derived from the word rec*ent*. They are shallow soils without a subsurface horizon. This very simple soil profile is due to one of the following reasons: (a) short time for regular soil evolution, e.g., on floodplains regularly flooded by muddy river water; (b) action of continuous erosion; (c) specific parent material like hard rocks, sands of deserts and semideserts, and mined lands; and (d) in mountains. They are nearly identical with WRB Regosols, and some of Entisols' great groups are similar to Leptosols. An example is the suborder Fluvents, great group Udifluvents, and subgroup Vertic Udifluvents.

Gelisols have the formation element *el*. The name has the roots in the Latin *gelare* which means to freeze. The parent material is weakly weathered and soil evolution is retarded or absolutely absent. Permafrost with typical honeycomb structure and lenses of ice occur in the entire profile. They are very similar to WRB Cryosols. An example is the suborder Histels, great group Folistels, and subgroup Lithic Folistels.

Histosols have the formation element *ist*. The name was derived from the Greek *histos* which means tissue. The topsoil is rich in organic matter; its content should be above 24 %. Their properties are either similar or close to WRB Histosols. An example is the suborder Folists, great group Ustifolists, and subgroup Lithic Ustifolists.

Inceptisols have the formation element *ept*. Their name was derived from the Latin *inceptus* which means start or beginning. They are soils with a weak development of horizons except for the top A horizon, or epipedon in the US terminology. These soils, usually young, display a soil profile that lacks the characteristics of other soils developed under the full assertion of soil-forming factors. They are nominated at their lower-level taxonomy as similar to certain WRB Gleysols, Stagnosols, and Umbrisols. An example is the suborder Ustepts, great group Calciustepts, and subgroup Udic Calciustepts.

Mollisols have the formation element *oll*. The name has its roots in the Latin *mollis* which means soft. They were developed in semiarid to subhumid climate where grasses dominated. Their profile does not indicate strong leaching – it could even be absent. They include the WRB Chernozems. And WRB Gleysols, Kastanozems, and Phaeozems appear among suborders of our Mollisols (see Ustolls). An example is the suborder Xerolls, great group Durixerolls, and subgroup Vertic Durixerolls.

Oxisols have the formation element *oxi*. The name is related to the term oxide. Very old soils belonging to this group were strong weathered in humid tropical climate with intensive leaching of Fe and Al oxides recognized by red color, especially in the illuvial horizon. They were earlier recognized as Latosols or Lateritic soils. WRB classification identifies them as Plinthosols and Ferralsols. An example is the suborder Ustox, great group Haplustox, and subgroup Plinthic Haplustox.

Spodosols have the formation element *od*. Their name was derived from the Greek *spodos* meaning ash. They are frequent in cold, moist climate with precipitations much higher than evaporation where the soil water flux is directed downward. Fe and Al oxides released from an acidic horizon are transported down to less acid horizon where they are accumulated. The leached eluviated horizon below the thin top horizon has a light gray to whitish color resembling the color of ash. The illuviated horizon has typical colors of rusty brown hues or dark brown. The original plants were coniferous trees and their needles contributed to the soil acidity. WRB classifies these soils as Podzols. An example is the suborder Orthods, great group Alorthods, and subgroup Arenic Ultic Alorthods.

Ultisols have the formation element *ult*. The name was derived from the Latin *ultimus* which means last. Evolving in humid tropical and subtropical climates that cause soil water saturation for portions of each year, their profiles contain mottled horizons and display other signs of waterlogging. They are identical with WRB Alisols, some Nitisols, while Ultisols' suborder Udults have all signs of WRB Acrisols. An example is the suborder Udults, great group Plinthudults, and subgroup Typic Plinthudults.

Vertisols have the formation element *ert*. The name has roots in the Latin *vertere* which means to turn. They are texturally heavy soils changing their volume with changes of water content. They are swelling when they are becoming water saturated, and they are consistently cracking each year during dry seasons. They have all the characteristic properties described in WRB Vertisols. An example is the suborder Xererts, great group Durixererts, and subgroup Halic Durixererts.

Soil faces differ from one location to another and are not uniform. We illustrate some of their unique, identifying features in the next dozen figures (Figs. 13.1, 13.2, 13.3, 13.4, 13.5, 13.6, 13.7, 13.8, 13.9, 13.10, 13.11, 13.12, and 13.13). The soil profiles demonstrated here reach roughly to a depth 1 m below the surface. The type of each soil is in the US classification at the level of order and suborder. In the World Reference Base (WRB), the Reference Soil Group (RSG) is indicated. If not indicated otherwise, the figures stem from the collections of Jan Nemecek, the Czech top pedologist. We are much obliged to him for his kindness.

Fig. 13.1 Alfisol, great group Hapludalf. At the top the umbric epipedon (humus A horizon) is nearly *black*, and below it is the gradual transition to the intensive *brown* illuviated B horizon which contains more clay particles leached out from the humus horizon. At the bottom is loess, the parent material (C horizon) with *whitish* thin $CaCO_3$ outcrops (WRB: Luvisol)

Fig. 13.2 Andisol, great group Haplidand. At the top umbric epipedon (humus A horizon) developed on volcanic ash, which forms the parent material (C horizon) (WRB: Andosol. Source: ISRIC, reference soil IT 016)

Fig. 13.3 Aridisol, suborder Orthid. The thin top horizon with a very low humus content contains soluble salts that effloresce on the topographic surface. Below it is the horizon with the highest content of soluble salts. Parent material is alluvium that is also usually saline (WRB: Solonchak)

Fig. 13.4 Aridisol, suborder Argid. At the top ochric epipedon (a thin humus A horizon) merges into an albic (*whitish*) eluvial E horizon that rests on a well-developed vertical columnar structure. Below it appears a natric horizon with menacingly high levels of exchangeable Na$^+$ (WRB: Solonetz)

Fig. 13.5 Entisol, suborder Psamment. A very thin ochric epipedon, difficult to ascertain as a recognizable A horizon owing primarily to its very low humus content, exists during its early stage of development. Unconsolidated sand remains at depths below the slowly developing A horizon (WRB: Regosol-Arenosol)

Fig. 13.6 Histosol, suborder Fibrist. At the *top* fibric epipedon with the dominance of undecomposed plants residues gradually transforms at greater depths into peat (WRB: Histosol)

Fig. 13.7 Inceptisol, suborder Aquept. Histic epipedon (humus A horizon consisting partly of peat) covers the albic horizon containing legendary signs of gleying with a transition to a mottled region that is regularly periodically waterlogged. The *bottom gray-blue* part of the profile is indicative of permanent waterlogging (WRB: Gleysol)

Fig. 13.8 Mollisol, suborder Ustoll. A deep mollic epipedon (humus A horizon) that gradually transforms into calcium-rich loess – the parent material (C horizon). WRB: Chernozem

Fig. 13.9 Oxisol, suborder Ustox. At the top the thin, imperfectly developed ochric epipedon (humus A horizon) lies over the oxic horizon containing secondary iron and aluminum oxides and hydroxides that give the typical *reddish* color and B horizon containing products of strong weathering developed to a depth of more than 1 m (WRB: Alisol)

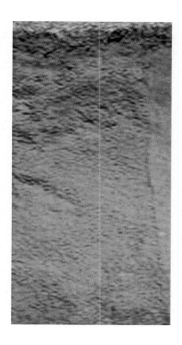

Fig. 13.10 Spodosol, suborder Orthod. Below the top ochric epipedon (a poor humus containing A horizon) lies the *white-gray* eluviated E horizon from which iron oxides were leached out yielding the *whitish ash*-like color and accumulating in the illuvial I horizon. The bottom is formed by transported weathered gneiss (WRB: Podzol)

Fig. 13.11 Ultisol, suborder Ustult. Below the ochric epipedon (humus A horizon) is the deep plinthite horizon of *reddish* color caused by free iron oxides that originated during long-lasting chemical weathering (WRB: Plinthosol. Source: en. wikipedia.org)

Fig. 13.12 Vertisol, suborder Udert. Deep cracks penetrate through the mollic epipedon (A horizon) to the alluvium sediment that forms the parent material (C horizon) (WRB: Vertisol. Source: USDA Soil Taxonomy)

13.3   Granny Soils

We accept and use the terms of units in time spans from geology to define and appropriately discuss the ancient lives of granny soils. Era is for time span of several hundred million years – there are 10 eras during the history of Earth. A shorter time span of tens of millions of years is named Epoch. Millions of years has the term Age. The term Period denotes a time span shorter than Era and larger than Epoch. Simplified: Era is for 10^8 years and more, Period for the range from 10^8 years to 10^7

Fig. 13.13 Vertisol's cracks
on the surface (Source:
Antonio Jordán, Spain)

years, Epoch for the range 10^7–10^6 years, and Age for 10^6 years and less except Holocene that started "only" 11.7×10^3 years before present.

We have seen that soil evolution was influenced by many factors, and among them one or two were usually dominant. In the US Soil Taxonomy, the climate took the dominant role in evolution of Alfisols, Aridisols, Gelisols, Mollisols and Oxisols and at least partly of Spodosols and Ultisols. We would like to mention here that soils were earlier classified using only two top taxons, climatic and aclimatic. This oversimplified scheme was abandoned many decades ago. Based on other soil-forming factors, organisms played an important role upon soil evolution in Alfisols, Mollisols, and Spodosols. Parent material and topography also had dominance in many orders.

Long before the shapes of present-day continents even existed, the climate, vegetation, and topography were all substantially and continually changing. Since they are also soil-forming factors, the soils originating in previous geological eras differed from our present-day soils. These old "granny" soils or paleosols are usually buried under younger sediments, or they are fossils within sedimentary sequences. Some remnants of old soils are hidden from the changing climate and vegetation below overhanging rock. Soils probably existed and changed in a very primitive form starting already 3,400 myr (3.4 billion years) BP, judging according to the existence of first sedimentary rocks and some fossil biochemical marks. The surface parts of igneous rocks were exposed to the influence of the atmosphere and they weathered. This material, let us say a "primitive soil," was eroded by rains and transported to its final destination – an ocean or a lake. There, muddy waters offered building material of new rocks by sedimentation of sand or silt and clay. Finally, the new, solid sedimentary rock was created by compression and cementation. Even though the origin of the "primitive soil" precedes the evolution of sedimentary rocks, we consider the first appearance of sedimentary rocks as a well-working approximation for the date of origin of the first "primitive soil." This estimated date could be slightly shifted to another value when future research rigorously identifies the origin of the first sedimentary rocks. However, soils in the context of today's meaning have existed only in the most recent eon, the Phanerozoic. More specifically, they appeared 454 myr (454 million years) ago with the action of living macro- and microorganisms, e.g., bacteria, and roots of plants.

Because the Earth is constantly changing, the program of its study could focus either on times of transitions, namely, transition from times with quasi-stable conditions to times with increased rate of change, or on the detection of soil properties in times of relative "quasi-stability." The study of paleosols in times of relative stability is preferable. Within this context paleosols could be studied as a natural body originating in the past and surviving either as a full soil body or as a vital step linking the past up to recent times. In some instances, only partial signs of the past soil are detected. Being mixed within recent soil properties, their identification requires sophisticated instrumentation applicable on the microscale or even on the nanoscale.

When the nineteenth-century geologists found remnants of fossil plants buried between sedimentary rocks having signs of horizontal fine strata, the hypothesis about old fossil soils was born. However, a conscientious study about them had to await the elaboration of soil science and soil evolution. Truly buried soils were first recognized in the nineteenth century between two layers of fair-colored loess each many meters thick and mutually separated by a thin humus soil horizon and eventually below it by a reddish-brown horizon. The top many meters thick layer of loess belonged to the last glacial era named Wisconsin in the USA or Würm in Central Europe. The bottom loess layer – below the thin buried humus horizon – originated in glacial era of Illinoian according to US terminology (230–130 kyr BP, kyr = thousand years) or named Riss in Europe. A substantial warming lasting for 15 kyr occurred between the time of Wisconsin and Illinoian glacials. This interglacial era is named Sangamonian in the USA and Eemian in Europe. The global, average scale of the climate warming was 3–5 °C higher than our recent Holocene global average. Owing to the warmer climate, the vegetation and other soil-forming factors also changed and are confirmed by the appearance of soil with a very distinct gray-black humus horizon. Above it the next Wisconsin (or Würm) loess (115–11.7 kyr BP) was deposited. Nothing more than the description of the buried soil hidden between two layers of loess was known at that time of the end of the nineteenth century in Russia and similarly in the midcontinental USA.

We identify the beginning of a new branch of science as well as a new scientific subject in the virtually unknown and forgotten booklet of K.D. Glinka, *The Aims of the Historical Pedology* (the title translated from Russian, published in Warsaw, 1904). The booklet offered a critical review of papers and books of geologists and geographers where the geological sediments were described in relation to plants and to rather incomplete knowledge of soils. Since *palaios* is Greek for old, the new branch of soil science was therefore named paleopedology and the broad family of old soils obtained the name paleosols. Glinka was probably one of the first scientists to be influenced by Dokuchaev who was a geographer and geologist. Glinka declared himself as a student of Dokuchaev and accentuated the relationships between the features of discovered paleosols and the impacts of climate, surface morphology, and plants responsible for their formation and evolution. Because he considered paleosols as reliable sources of information about climate and universal soil-forming factors at the time of their origin, Glinka was ahead of his contemporaries by more than a quarter of the century. His ability to understand broad, universal links encouraged and enhanced cooperation with soil scientists abroad and accounts for him being one of the cofounders of the International Soil Science Society in Rome in 1924.

One relationship that has proven useful for paleosols is that between the depth to the calcic horizon and the mean annual rainfall. A new compilation of data presented here demonstrates that this relationship holds throughout the world for arid land soils. The use of this relationship for interpreting the paleoclimate from present-day observations of paleosols is easily illustrated with an extinction event occurring 33.9 myr (million years) ago during the transition from the Eocene to the Oligocene. By the way, it was not a truly extreme event in the Earth's history even if it is called Grande Coupure (Great Break). Cetaceans (such as a whale, dolphin, or porpoise) were the main affected animals. The general cooling was accompanied by climate phases documented by transitions from Entisols to Inceptisols and by the upward shifting of calcic horizons.

Estimating the age of observed relics of past soils is a very important part of studying paleosols. For the youngest paleosols, which evolved not earlier than 40,000 years ago, radiocarbon dating is commonly used. It is based upon the determination of the ratio of "ordinary" carbon ^{12}C to radioactive carbon ^{14}C. With the top index denoting the number of neutrons in the carbon nucleus, we know that ^{12}C has 12 neutrons and is stable. We also know that ^{14}C is not stable since it has 2 additional neutrons and is in a continual transition to eventually become stable, nonradioactive nitrogen ^{14}N. Because the rate of ^{14}C decay is in equilibrium with the rate of appearance of new ^{14}C created by cosmic radiation, the relative concentration $^{12}C/^{14}C$ in the atmosphere is virtually constant. We say virtually constant owing to the fact that minute oscillations of cosmic radiation occur. However, this slight oscillation around its average value is negligible in our dating method. We are taking 1 atom of ^{14}C as related to 1 trillion atoms of ^{12}C. The same ratio of those two isotopes exists also in living organisms, since this constant ratio enters into plants due to photosynthesis. When atmospheric carbon forms organic molecules, the ratio of the two isotopes is kept the same in the new organic molecule like in the atmosphere and the same is valid for animals consuming the plants and the same happens to plankton in waters. All living bodies on the Earth, including us, *Homo sapiens*, have the concentration $^{12}C/^{14}C$ in equilibrium with that in the atmosphere. The decayed atoms ^{14}C in living bodies are replaced by the ^{14}C from the atmosphere. At the moment of death of the animal or the plant, all metabolic functions stop and the acceptance of atmospheric mixture $^{12}C/^{14}C$ also stops. Because ^{14}C is not stable, a slow decay is running and the atom ^{14}C is transformed into ^{14}N and a weak beta radiation inside of the dead body. However, this decayed isotope ^{14}C inside of the organic molecule is not replaced by the new isotope from the atmosphere since the organism is dead and the earlier living functions have stopped. The consequence of the no more existing reaction between the plant (or the animal) and the atmosphere is a gradual decrease of ^{14}C concentration in the dead body. One-half of the ^{14}C atoms decay after 5,730 years – the time period denoted as its half-life. Using its half-life value in an exponential equation, we obtain the time of death of the body being analyzed in our study. The accuracy of the analytical method allows the age of the body to be estimated for times not more than roughly 60,000 years BP (before present) with the year 1950 taken as the present time. Most frequently, the method is used for estimating the age of unconsolidated sediments, shells, some limestones, and organic materials. According to the half-life of ^{14}C, the radiocarbon method is applicable back to the middle of the last glacial period, denoted as Würm in the Central and Alpine Europe, or Wisconsin in the USA, or Weichselian in the north Europe, or Merida in Venezuela.

The age of older Quaternary sediments and buried soils has been determined by optical methods which translate the time of stored luminescence into age. Although several methods have been developed, the most frequently used procedure is that called optically stimulated luminescence (OSL). Exposure to sunlight resets the luminescent signal and so the time period between recent time and the time when the soil was buried can be calculated. The methods are applicable for ages between 500 and 100,000 years.

Determination of the age of "very young" paleosols developed in Holocene requires more accurate methods. For example, the spores of mycorrhizae are used for identifying buried remnants of A horizon below a recent soil or of its top horizon when a relatively small age up to 5,000 years BP is the object of the research.

In reports on sediments and soils in the time span between recent and about 170,000 years BP, a detailed time scale is described by the term MIS (Marine Isotope Stage), and in less frequent time named Oxygen Isotope Stage by some authors. It is based upon the ratio of oxygen isotopes, $\delta^{18}O$, or ratio of "heavy" oxygen to "light" oxygen ($^{18}O/^{16}O$) in studies of ocean bottom sediments sampled by core drilling. Up to now, over 100 stages have been identified reaching slightly over 6 million years before present, i.e., the stages cover the youngest part of Miocene, Pliocene, Pleistocene, and Holocene. The cycles in the isotope ratio correspond to glacials and interglacials. Even-numbered stages have high levels of oxygen-18 and represent cold glacial phases, while the stages with odd numbers have a relatively low content of oxygen-18 and they represent warm interglacial intervals. MIS 1 is for Holocene, ending with Younger Dryas. The last glacial (Wisconsin or Würm or Weichsel glacial) contains MIS 2–5. Some stages are subdivided into substages, like MIS 5, where 5a, 5c, and 5e denote relatively warm substages and 5b and 5d are cold substages. Substage 5d is the end of last interglacial (115 kyr BP) and starts the last glacial, while MIS 5e denotes part of the last interglacial. The numeric values continue into a deeper past, and, e.g., the end of the Pliocene is denoted by MIS 103 (2.588 myr BP).

The estimation of a larger time BP is also performed on the principle of stable isotopes being related to the unstable isotope of the same element. Relatively frequent is the use of potassium isotope method. The name of the element in English was derived from "pot ash" that refers to the method of obtaining potash during and before the Middle Ages – that is, burning wood, watering the product of burning, and next evaporating the solution in a pot. The Neo-Latin kalium has its roots in the word alkali which was derived from the Arabic *al-qalyah* that also means plant ashes. The stable, mainly occurring ^{39}K is related to the non-stable isotope ^{40}K with a half-life of 1.25 billion years that decays to stable argon ^{40}Ar. The ratio of these two K isotopes applied for a very long time between 100,000 and 4 billion years BP is generally used for estimating the age of biotite and volcanic rocks. The use of uranium isotopes does not strictly fit the method derived for carbon dating because all three natural isotopes of uranium are unstable and decay differently. Although several alternate procedures have been developed for uranium age dating, the application of ^{238}U/^{206}Pb with a half-life 4.5 billion years is decisive and is applicable for ascertaining ages for times less than the half-life. Reliable dating was successfully achieved. Rubidium and strontium isotopes ^{87}Rb/^{87}Sr provide age dating opportunities, too, similar to isotopes ^{85}Kr and ^{3}H/^{3}He.

With paleosols never being found as undisturbed soil profiles, researchers generally discover only small parts of the original profile, or on rare occasions when they have good luck, they find intact segments of the original horizons. Usually they have to study micromorphological bodies in the complete absence of macroaggregates. The main task is to describe the shape of microaggregates and the material of globular components similar to microspheres. Additionally, the shape and material of thin films covering silt and sand particles, the cutans, are described. The name for these microforms, introduced in 1964 by Roy Brewer, was accepted by soil scientists. Based on fabric and shape of cutans, local conditions of soil evolution are classified. The main grouping of cutan fabric into major categories is as follows: carbonates, iron oxides (sesquans), organic matter (organans), and type of clay minerals (argillans). The kind of solid surface of silt and sand also plays an important role in the formation of the shape of cutans. In addition to particle cutans, there are ped cutans covering the surface of microporous peds and void cutans filling the micro-voids where the particles contact each other. With cutans being further subdivided into products of eluviation, illuviation, diffusion, and stress action, details of soil processes can be identified. Through them, the soil type is detected as it was evolved in the past. Occasionally, other remnants of weathering could also be found.

For example, if kaolinites prevail in clays of cutans of the soil protected now from recent climate change and vegetation, soil paleontologists will search for iron concretions. And if they find them, they will tentatively accept the hypothesis of Latosol evolution during the past era and will try to confirm their assumption by other evidence. Finding paleopedologic evidence that the climate was warmer and more humid in that specific era, they verify and accept their hypothesis of Latosol evolution. When there are residues of Vertisol evolution characterized by traces of distinct swelling in rainy seasons and shrinkage during rainless seasons, the paleosoil scientist will study the thin films on micro-cracks within soil microsections. The cutan coating of microaggregates is similarly studied. Because this type of research requires observations at the microscopic or even nanoscopic level, studies were performed only recently using contemporary fine instrumentation that enabled chemical and shape analysis to be realized on ultramicroscopic size of soil samples.

Magnetic properties are important indicators about the quality of past soil development. Great magnetic enhancement was estimated in the top horizon of well-drained paleo-cambisols where weathering had contributed to the concentration of magnetic silt-sized fine grains and cutans. Generally, the well-drained and intermittent wet/dry soils manifest enhanced magnetic properties on the ultra-microscale even a hundred thousand years after their genesis. On the other hand, the excessive arid or wet and acid soils were not able to form significant amounts of micro-ferrimagnets in paleosols. The initial stages of the weathering of loess can also be identified by changing concentrations of micro-ferrimagnets.

Soluble silica absorbed by plant roots from the soil is carried up through the plant to some of its organs, and after the plant or its parts die and break down, this "skeleton" – usually of microscopic size – remains in the soil. Its scientific name is phytolith, derived from the Greek *phyton* meaning plant and *lithos* meaning stone. When phytoliths are found and used for the detection of the type of vegetation, the paleo-soil scientist has in hand one of the important soil-forming factors that can be used to confirm or reject a hypothesis regarding the genesis and existence of an actual soil during a specific geological era. The phytoliths are hidden primarily under old river sediments (alluvium) or loess layers that originated in glacial/interglacial cycles during the last 2.588 million years (Pleistocene). In some instances, residues of paleosols are under the colluvium splashed downslope, or their parts are hidden below a layer of volcanic ash. Generally, the occurrence of a complete or substantial part of an entire soil profile is a rarity. Paleo-soil scientists and geologists usually find only microscopic to small traces of past soils. Each such discovery has helped us to separate individual glacials since if two layers are separated by traces of soil, the meaning is simple – a glacial there could not exist because soil evolution is impossible in the presence of ice. In a similar way, volcanic ash sedimentation was interrupted by a time period when the ash was a parent material of new developing soil and the volcanic activity stopped for a time period of sufficient length for soil development. Generally, we could also compare results from numerous other methods to detect memoirs of paleosols.

When we know the age and type of soil in a certain region, we can change our research upside down and own a tool for estimating the climate during times of paleosol evolution. Paleo-soil scientists found the remnants of Terra Rossa in the Rendzina region of Central Europe. The climate typical for the evolution of Terra Rossa was 2–3 °C higher than the recent annual average temperatures of contemporary Rendzina. Moreover, there were dry seasons lasting each year for several months – a condition required for the evolution of Terra Rossa in the time of the last interglacial Sangamonian (Eemian) 130,000–115,000 years BP. A couple of other paleosols originating in the last interglacial further document that the climate was warmer than in our recent interglacial Holocene.

Layers of sand in several regions of present-day deserts cover thick crusts containing $CaCO_3$ – a reliable sign of the past existence of soils across large areas. Somewhere there must be remnants of dark gray A horizons indicative of a less arid climate with rains coming regularly each year. We have proxies about the existence of either low or even high savanna with lots of animals in significant regions of recent Sahara desert or in the large Indian desert Thar running along the border between India and Pakistan. Isotope studies of layers of the Thar Desert show wet periods during the last two interglacials. Having mentioned the decrease of the areal extent of deserts according to paleosol studies, we also note increased areas of desert. Deep drilling in all recent deserts and semideserts shows that the extent of deserts was largest about 25,000–20,000 years ago during the coolest period of the last Wisconsin (or Würm) glacial. During this time period, based primarily on carbon isotope studies, this cool "peak" occurred in the last glacial with the extreme global dryness.

Importance of the study of paleosols was discovered in agriculture and ecology, since the knowledge of local remnants of paleosol enables us to predict the next change of soil properties due to eventual introduction of new plant types and cultivation techniques, or under the influence of industrial and generally technological activity. We have to keep in mind that ongoing increasing or decreasing changes of contemporary soil fertility cannot be accurately predicted based on only a few years of measurements or even those observed for a sequence of several decades. To truly understand the reality of seasonal local weather coupled with global climate demands that we deal with scales that connect decades with hundreds and thousands of years. Such changes would be irreversible under natural conditions, and if the change was in the direction of decreased soil fertility, the amelioration would be very expensive, even if possible at all. Briefly saying, the knowledge of the soil history enables us to understand the recent and future changes.

13.4   Maps of the Soil Distribution

There are two types of soil maps. In a simpler or local case, the map covers a relatively small area with a scale in ranges roughly 1:100 to 1:1,000. It is frequently ordered by the farmer or by the consortium of several farmers, or by the authority of a local or regional district. Sometimes when farmers or owners of land need more specific information than is normally displayed on an ordinary small-scale soil map, they request a soil cartogram that illustrates the geographical presentation of a certain soil property – like soil pH, soil texture, depth of A horizon, quantities of available plant nutrients, etc. The elaboration of such a map requires that a soil scientist has an effective combination of brain and brawn, brawn because strong, agile legs are needed to walk through the landscape while looking for appropriate locations into which soil pits are next dug with a shovel using the strength of powerful biceps. His brain decides the initial number and spacing of the pits. If his brawn is deficient or if he gets too tired to dig all of the pits, he depends on younger colleagues to finish the digging. After all of the pits are completely dug out, the soil scientist walks back and forth between each and every pit several times in order to personally make detailed observations and measurements within each profile and to remove large numbers of soil samples that must be analyzed in the laboratory. His brain takes the next step to identify the geographical distribution of specific soil properties based on a summary of his knowledge of the soil types, terrain, and vegetation at the location of each pit. If the soil types or mapped properties in two neighboring pits differ, he drills several holes between the two pits using a hand- or motor-driven soil sampler in order to more precisely determine the mutual boundary. And thus, with that level of sampling, the first sketch of a soil map across the terrain becomes a reality.

A completely different technique is used when a soil map of a large country, continent, or even of our planet Earth is ordered. Cooperating with institutions and commissions of specialists, the soil scientist sits in his office while collecting maps and all relevant materials of the region being mapped. A general rule prevails – the smaller is the scale, the larger is the mapped area, number of institutions and commissions, and number of cooperating specialists. It is important therefore that the head soil scientist including the entire ensemble of coauthors has rich experiences in mapping local soils and, if possible, those soils in different regions exposed to various factors like climate, vegetation, land use, and soil technology. Hence, to develop a map across large landscapes, a different combination of brawn and brain is required. Instead of strong, healthy legs, it is far more important to have vigorous sitting muscles associated with a tireless brain to allow the soil scientist to continuously and conscientiously participate without exhaustion during uncountable sequences of meetings.

Chapter 14
Illness and Death of Soils

14.1   Can the Soil Be Ill?

The ability of soils to produce more and more foods for mankind and for animals kept by mankind is not limitless, and the increase rate of this production appears slower than the rise of the number of people on our planet Earth. When the two types of increase are plotted versus future time, the two curves shall eventually cross each other. See Fig. 14.1.

Additionally, our humane aspects say that we have to decrease substantially the percentage of undernourished people, and this requirement reduces the critical time when the two curves cross each other. Our aim should be to shift the time of crossing as far to the future as possible, while the percentage of undernourished and starving people would sink to zero.

Immediately, we must ask a primary question: Are our recent attempts – or more appropriately speaking, attempts of politicians – to produce and harvest plants for manufacturing a portion of green energy really rational, essential, and prosperous for mankind? With our cozy and cushy lifestyle requiring a strong rise of energy production, a new branch of industry appeared, named green energy that we as authors prefer to call alternative energy. The word green could be misunderstood and thought to be identical with using plants as a source of energy production called biofuels. However, several industry experts believe that by 2050, biofuels could provide at least 25–35 % of the world's transport fuels. The technology is based partly on direct burning of crop wastes and partly on biomass gasification. It means that sugar, traditionally used as a constituent of food, is converted to biodiesel. Other types of conversion processes use yeast fermentation, bacteria for producing biodiesel from cellulosic materials, and algae as a potential biofuel feedstock.

The production of biofuels leads the alternative or so-called green energy industry into a blind alley. The misuse of soil as a medium in energy production instead of that for feeding mankind is equivalent to a soil illness brought into the natural process as an infection by irresponsible technologists and politicians. Their recent

© Springer Science+Business Media Dordrecht 2015
M. Kutílek, D.R. Nielsen, *Soil: The Skin of the Planet Earth*,
DOI 10.1007/978-94-017-9789-4_14

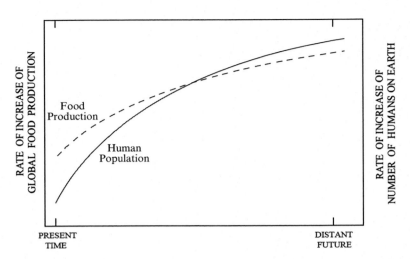

Fig. 14.1 Rate of increase of potential food production against the time and rate of increase of population against the time

blatant public campaign in favor of biofuels leads the soil to deviate from its principal role in the environment, which could be defined briefly as feeding plants and thus directly feeding herbivores and finally indirectly feeding carnivores including *Homo sapiens*. If this principal role of soils is omitted due to the preference of production of biofuels, our basic question is: What is more important for us as a species? Should we avoid famines of mankind or should we extend a lazy life full of comfort to a small surviving percentage?

Additionally, let us consider the third rational reality of gaining more energy by exploiting the energy hidden in chemical reactions on the atomic and molecular level of the inorganic world. The deviation of principal aims of soil in the environment, taking into consideration our human views, is similar to the action of some viruses which attack a certain part of an organism and cause its inactivity in the frame of the other neighboring parts. Subsequently, the entire organism suffers and eventually dies. The propaganda and political and economic support of biofuel production are therefore very similar to infection by viruses. However, the human society is still strong enough to stop this infection. We are not discussing a horrifying terminological fault, as if coal, oil, and gas were not products of processes transforming the original parental living organisms like trees to coal or single cells of plankton into methane gas and petroleum (liquid and gaseous hydrocarbons). By the way, the literal translation of petroleum is "rock oil," and it was introduced by Georgius Agricola in the sixteenth century as new word stemming from Latin *petra* meaning rock and *oleum* meaning oil or the Greek *elaion*.

Not being linguists, let us return to Fig. 14.1. The aim of shifting the crossing point as far to the future as possible can be reached in principal by two ways. One would be the reduction of the rate of increase of the number of humans on Earth. Expressed in simple language, it would mean the limitation of births. It is simply written, but the ways of realization are so complicated, it does not look realistic. Indeed, whenever and whoever tried to even start with such limits, the result was catastrophic. Although we are not speaking about ethics, we are not qualified to enter this mined field. The other way to shift the crossing point further into the future is to increase our care and enhance our abilities to keep soils in a "healthy" state. In other words, our duty is to fight against various soil illnesses and try to prevent their appearance. However, if the principle of soil sustainability is accepted, why do we ever so frequently introduce the magnitude of yields as the criterion for optimal soil management? And here a new problem or question is emerging.

When we speak generally about the need to maintain biodiversity on behalf of sustainability, why has there been no mention of taking steps to preserve pedodiversity? Present-day, repetitious activities are directed to transform naturally occurring diverse soils into a bastard domain across a field using artificially introduced procedures directed to achieve those properties approaching characteristics of fertile soils developed under natural conditions. Is it not the time to now define the boundaries of spatial domains that limit our practical successful activity that bastardizes the natural diversity of soils within farmers' fields? As authors, we did try to find an appropriate definition of the problem without speaking about its satisfactory solution. But independent of our attempts to answer the emerged question, the need to fight against soil illnesses remains.

We have described one of several illnesses in our earlier chapters. It was and still is the deterioration of soil structure and sometimes even a complete loss of the structure. It is related to the change of the original vegetation that was destroyed and new plants introduced at the start of agriculture – the greatest revolution of mankind. The major part of agricultural plants is regularly harvested and removed from cultivated fields. The portion of plants remaining in each field does not cover the loss of humus due to its decay. Even the most resistant humus substances are decaying with a half-life of several years up to about two decades. Whenever there is not enough "raw organic matter" being returned to the soil and transformed into humus, the total content of humus in soil decreases substantially within two or three decades that severely degrades the soil's ability to adequately support its vegetation. This is why new terms have been coined as anthropodegradation of soil structure or anthropo-conversion of land forms and other terms starting with the Greek *anthropos* meaning human being. The invention of new words having glimmers of science did not solve the problem. It was already known that the loss of humus and especially of glomalin is one of the most dangerous signs of soil deterioration, or let us say of soil illness. The contemporaneous trend aimed at preventing this illness is to use procedures that retain maximum amounts of plant remnants in the field to provide supplies of organic substances readily available for beneficial, continual humification.

The improvement of agricultural methods brought about another technique. It was plowing that evolved from the primitive loosening of soil by a simple hoeing. The hoe was gradually transformed into a soil-loosening scratch plow when animal traction was invented. The new instrument was a vertical wooden stick fixed to a wooden frame that was pulled by a draft ox. The wooden stick broke up a narrow strip of land along the direction of this "plowing" line. Since the method left a strip of untouched soil between the neighboring rows, the soil was frequently cross loosened by mutually perpendicular paths. The ox-improved tractive force, later improved with a horse and sometimes vice versa, was also used with the vertical stick being replaced by a moldboard. At the end of this period, wheels replaced the runner and the compressive role of the moldboard plow weight was reduced. Earlier, the wheelless plow had to be dragged across the field. The shape of the moldboard plow aided horizontal cutting as well as turning the loosened topsoil layer upside down so that its original surface lay on the bottom of the plowed trench and its base became the new soil surface. Finally, the tractor replaced the horse and the depth of tillage was increased. Consequently, plant nutrients and humus in lower portions of the A horizon reach the soil surface, while the earlier impoverished surface soil is shifted deeper into the profile. Because the positive effects of repeated tillage to the same depth steadily decreased, it was necessary to plow the soil to greater depths to renew those positive features of churning the A horizon. However, greater depths brought higher friction from an increased soil resistance that was overcome by using a stronger tractive force. Horses were replaced by tractors. And, the stronger the tractor, the deeper the soil could be plowed. On the other hand, the increased weight of stronger tractors progressively compressed the soil causing a loss of the structural stability of the original loosely packed aggregates of friable soil. During compression, individual soil particles are pushed together reducing pore space and increasing the soil bulk density. With the coarse pores being wiped out, the porosity is greatly reduced. We must also remember that this physically created condition is more intensive when there has been a loss of soil humus, especially of humins and glomalin, that accelerates the disintegration of soil aggregates.

Soil compression is most apparent within the A horizon. But the top part of the horizon is loosened to a maximal porosity during plowing to prepare the soil surface for the next crop. After planting, during the vegetative season and after harvest, the porosity continually diminishes until the soil is once again plowed and prepared for another crop. However, the layer below the plowing depth is permanently compressed, and after a long time, a distinct permanently compressed horizon (plow sublayer) is formed. The regularly "loosened" top part of the A horizon also retains marks of compression for a long time, but they are less impressive than those observed below the depth of plowing in the plow sublayer. Plowing is also a very effective tool for burying weeds. Its effectivity is at least as powerful as the contemporary use of herbicides. Plowing has the great advantage that it does not endanger groundwaters and surface waters in lakes by herbicides or their derivatives after being chemically transformed in soil and in surface waters.

Soil compression also has unfavorable consequences. The rooting of plants is worsened, and when the roots of agricultural plants reach the depth of the permanently compressed plow sublayer, they are not apt to penetrate through it. Consequently, the plants start to suffer from a lack of nutrients. The reduced porosity brings reduced soil permeability, and the lower is the permeability (or hydraulic conductivity), the smaller is the portion of rainwater penetrating the profile. Most of the rainwater attacks the sloppy soil surface causing erosion and eventually the potential collapse of the soil vegetation owing to incoherent environmental conditions. Water availability becomes an issue – the plants suffer from droughts during extended rainless periods, or they suffer from lack of water during shorter rainless periods. Or when there are adequate levels of plant available water within the profile, owing to the absence of coarse pores in the compressed A horizon and plow sublayer, the exchange of air between the soil and atmosphere is severely restricted. And with a lot of oxygen in the soil air being consumed by the decomposition of organic compounds, the soil could even suffer from anaerobic conditions where the transformation processes produce toxic compounds and the soil increases its acidity.

Reclamation of the soil at the top of the A horizon is not simple. The compressed surface without stabile structure forms a crust when soil dries out. When it is wetted by rain, the surface is transformed into a muddy puddle surface not readily permeable to either air or rainwater or to melting snow at the end of a cool winter. All these changes lead to worsening the condition for planting agricultural crops and to the reduction of their yields harvested at the end of their growing season.

There is a lot more than just plants that live and grow in soil. Soil-inhabiting organisms have a variety of types and sizes, e.g., small burrowing mammals, insects, amphibians, and worms seen with our eyes, and microbiota like nematodes, bacteria, fungi, and actinomycetes. All of them work to decompose, transform, and disperse plant and soil organic residues. They also in many instances help to weather soil minerals and to enable the dispersion of less soluble compounds. All those actions cause the release of plant nutrients in mineral forms available for the initial growth of new plants. Each and every process in soil is a decisive part of nutrient cycling. Although nutrients from soil organic residues in forms of inorganic compounds are initially not available to plants, they are gradually transformed, stored in soils, not leached out, and gradually released at times related to the specific needs of cultivated plants. The intensive decomposition of organic compounds and release of nutrients are realized provided that the soil has an adequate quality.

Another type of soil illness is salinization. Often occurring naturally in arid and semiarid zones, it was accelerated by irrigation when agriculture started. Without irrigation, man's greatest revolution could not exist in such zones. Simple irrigation was merely an imitation of nature, of floods across a natural landscape where the soil is fully saturated by an excess of water. An excess of water means the leaching of readily soluble salts that were released during the chemical weathering of rocks or geological sediments. Since the dominant net flow of water is in the upward direction, the solutes are also transported upward as water evaporates from the soil surface. Each simple irrigation or pulse of soil saturation by water has the consequence of transporting more salt to the topographical surface. Likewise, when the groundwater level rises owing to the seepage of water from the irrigation ditches and channels, both water and its dissolved salts migrate upward toward and into the root zone. The elevated salt concentration in the root zone does not allow economical farming. As harvests drop because the soil becomes progressively less fertile, the soil is ill according to our criteria. The best way of curing it is the prevention of salinization. Local governments should avoid wasting water by seepage from poorly designed and installed irrigation structures. They should avoid salinization of surface waters by industrial discharge of liquid wastes and by salts leached from upstream fields. In some special large regions, the farmers cooperatively organized and carried out regular or even alternate-year fallowing. It enabled the water table to fall after each harvest due to evaporation and transpiration from the wild plants that took over the cultivated fields when irrigation was interrupted. However, whenever soils are already saline, their amelioration is expensive and necessitates a sequence of several steps of washing and draining an excess of water completely through the root zone. The main problem is where to conduct and dispose of the saline drainage water. Because downstream governments are usually against transferring salty water into one or more nearby upstream riverbeds, the curing of soils suffering salinization is complicated and expensive. Inasmuch as salinity reduces yields on about 20 % of all irrigated soils, its impact is certainly not negligible considering that about 12 % of the global population is still undernourished (FAO statistics for 2010–2012).

If we say that soil is something like a living organism, then we must admit that the degradation and worsening of soil properties are similar to a serious illness of the soil. Moreover, an ill or sick soil equals starving, and starving means losing battles and finally the loss of a war. All such circumstances have been frequently caused by the abrupt change of some of the natural conditions in the past, but we realize additionally that various activities of human society also contributed. And no doubt, all shall contribute in the future. Our aim is to understand how soil is the decisive factor in the environment.

14.2   The Death of Soils

Water is muddy during a flood, and when the high water sinks back into the riverbed, the surrounding fields and meadows are covered by a film of sludge – a dark gray, flimsy almost transparent shroud extending over and between thin branches of shrubs, blades, and straw as well as the soil surface. The constituents within these shrouds originated upstream, were carried along in the flood, and are no longer present upstream. What is really missing upstream? The answer is surface soil eroded by water from the place of its origin. It is as if the soil dies. Rainwater flowing on the slope dislodges and carries soil particles down the slope. Great volumes of this concoction of water with silt and clay particles are transported to the plain below the slope and eventually into the mainstream of a brook and river.

Soil water erosion is caused by water flowing in a thin sheet on the slope of land being attacked by raindrops. When the rain intensity is higher than the infiltration rate, the excess water remaining on the land surface forms a very shallow water layer having a very small slope that causes a downhill flux of this excess water. The falling water drops look like a strong bombardment, the soil aggregates are mechanically broken into elementary silt and clay particles, and the released particles are smashed around. These elementary particles are carried away even by a small flux of surface water because they are each very tiny and substantially lighter than the heavier aggregated particles. The existence of this process is sometimes called *splash erosion*. The easier the aggregates are disintegrated, the greater is the mass of fine soil particles splattered into the water and transported downslope. The longer is the slope, the more intensive is soil erosion, because the thickness of the sheet of flowing water increases as well as its carrying force and momentum. A similar relationship exists between the angle or steepness of the slope and intensity of erosion. When we observe soil profiles along the wall of a trench dug in the downhill direction of an eroded soil, we see that the humus horizon thickness decreases as we go down the slope. The top, missing portion of the humus horizon was carried away by *sheet erosion*. Although the eroded soil may not move entirely off the field during the first signs of water erosion, it is moved from higher to lower elevations, causing some parts of the field to be less productive. In reality, there does not exist a regular sheetlike flow of the muddy water. The majority of water flowing downslope readily forms initial irregularities owing to the roughness of the existing initial topographical surface. Hence, very small crevices and miniature gutters are dug out by water streaming across naturally created surfaces. These small ridgelike miniature formations are smoothed after subsequent rain by both natural factors and man-made tools. If they survive until the next rain, they will gravitationally attract the excess water from the rain. This type of migrating surface flow is entirely natural and contains no magic. It is caused by the same principles that led to the first formation of small channels or rills. But since the surface flux is now more directed into the connecting formation of rills, water readily flows into the accessible rills ahead of it and does not consume its energy searching for an optimal path down the slope. Following paths already predetermined down and across the field, water's increasing and conserved energy is released for the intensive transport of soil particles. Hence, the initially small, fragile rills sooner or later become deeper and broader and survive over many years as the result of *rill erosion*. The rills may be interconnected into a sort of a net covering the whole field.

The more frequent surface runoff is conducted by a rill, the higher is the probability that soil particles erode from the sides and bottom of the rill until the rill is made so deep that we speak about a gully. It reaches deep below the humus horizon of the surrounding terrain, and the process is denoted as *gully erosion*. Gully erosion together with rill erosion destroys the earlier existing soil, and if the erosion continues for a long time, the landscape is denuded of soil and is no longer capable of supporting plants. Hence, the soil dies and loses all of its ability to support life – plants, animals, and humans. Therefore, we have the priority and commitment to minimize soil erosion to an estimated designated tolerable soil loss level TSL. By universal global agreement, it has been determined that TSL should at least equal to the rate of contemporary soil evolution. Hence, its value roughly ranges between 0.1 and 1 ton per acre per year, or equivalent to 0.1–1 mm of topsoil layer in 1 year.

Soil may also be shifted from the field by another transporter, the prevailing wind. The process, denoted as *wind erosion*, can happen anywhere and at any time when the wind blows with a sufficient wind speed (velocity) measured at the height of 10 m above the ground at a standard meteorological station. In meteorology, wind is classified with the Beaufort wind force scale. Here we use its simple, more convenient name – the Beaufort scale. Although measured in wind speed units, not force units, the numerical values of the scale are related to empirical observations. The wind is classified as *breeze* (light, gentle, moderate, fresh, and strong) up to Beaufort number 6 (maximum 49 km/h). *Gale* starts at Beaufort number 7 as moderate and ends at 9 as strong (up to 88 km/h). *Storm* is classified at numbers 10 and 11. The number 12 and eventually higher is denoted as *hurricane* (for 118 km/h and more).

The wind dries out the soil surface, and as water is lost from the soil aggregates, the wind disintegrates the dry aggregates into elementary particles. The silt and clay particles are then easily lifted by the turbulently blowing wind that is especially strong when there is sparse vegetation. In arid zones where the land is virtually without any vegetation, the lift of particles is extremely easy. Once the tiny particles are in the air above the soil surface, they are simply carried away by the wind. If the elementary particles are only weakly bonded together, their quasi-aggregates are easily broken by the force of wind, and the air erosion is more vigorous. Not only does wind damage the land by erosion, but the sedimentation of the dust particles, usually at great distance, causes damage not only on highways but to vegetation since the plants are suffocated by their stomata being plugged by clay and silt. And when the process occurs in hot weather, the vegetation is even burned by hot particles. The higher are both wind speed and air temperature, the more intensive is wind erosion with completely negative consequences. The higher the wind speed is and the air temperature is higher as well, the more intensive is the wind erosion with all negative consequences. In many dry areas around the world, when wind erosion gradually converts fertile land into wasteland, we can truthfully speak about the death of soil.

The process of wind erosion is recognized in three phases:

1. Movement of soil particles smaller than 0.1 mm suspended in the air. The suspension is moved up to height above the land surface, and there, parallel to land surface, the particles are carried far away like a large cloud. They return to earth only when the wind subsides or when there is a high obstacle oriented crosswise to the blowing wind. It is similar to the formation of a snowdrift. If the process is repeated many times during thousands of years, a new layer of sediments is formed. Our best example of such erosion and sedimentation is the 10-m thick loess layers that originated from a product of wind erosion during the glacial periods in the last 500,000 years of our recent Pleistocene. On the other hand, if we observe only one erosion phase, we sometimes discover that the wind erosion stops when the particles are carried downward with precipitation.

2. Movement of particles by a series of short bounces along the surface of the ground and dislodging additional particles with each impact. The process is denoted as saltation. The bouncing particles ranging in size from 0.1 to 0.5 mm usually remain within 30 cm of the soil surface. Depending on conditions, this process accounts for 50–90 % of the total movement of soil by wind.

3. The rolling and sliding of larger soil particles along the ground surface, known as soil creep. The movement of these particles is aided by the bouncing impacts

of the saltating particles described above. Soil creep can move particles ranging from 0.5 to 1 mm in diameter and accounts for 5–25 % of total soil movement by wind.

There are several ways of controlling wind erosion. The cohesion of the constituents inside of aggregates is increased when the content of soil humus is increased and when the humification leads to the increase of glomalin content. If the roughness of land surface is increased, the wind speed just above the surface is reduced. A dense vegetation acts in a similar way to reduce the wind speed in the air close to the land surface. Rows of trees around a field serve as a windbreak to protect the enclosed area against wind erosion.

14.3   How to Keep the Soil Healthy

The first step is directed to the increase of aggregate stability and to balanced plant nutrient resources. We have already mentioned that the majority of nutrients in soil are stored in forms that are not immediately available to plants. At first thought, this seems ineffective, but it is a necessary feature of soils since only such nutrients are stored and not leached out of soils. They are released as a function of time and usually in an adequate manner. Plant roots exude chemicals that help dissolve some of the less soluble compounds as plant nutrients are being released. Plants are not able to achieve this release completely by themselves.

Other organisms inhabiting soils help out and among them are the very effective and beneficial fungi called mycorrhizae (M). They have the ability to grow into the roots of many plants, and at the same time, their fungal hyphae or branches grow into the soil matrix. M fungi live symbiotically with the crops because they allow access of needed nutrients from the soil to their plant partner. And then it is the plant's turn to offer and share with the fungi its photosynthetically produced sugars as an energy source. The M fungi make nutrients more accessible to crop plants by several processes. They increase the extent of the plant root system within the soil by exploring additional portions of soil favorable for root growth that enables a better touch between the root and the soil matrix. Moreover, the M fungi have the ability to dissolve slightly soluble compounds containing plant nutrients. This ability is especially important during instances when the exudation of crop roots is inadequate or less effective. Many plant nutrients even in soluble concentrations have low mobility in soil. They are able to diffuse slowly and only few millimeters from their release to the consumer – the plant root. It is therefore very efficient when nutrients are consumed very close to their release from more complex insoluble compounds by M fungi. The effectiveness of M fungi can be easily increased by choosing specific crop rotations and reduced tillage systems. This principle is especially important for the less mobile macronutrients of phosphorus and potassium and the majority of most micronutrients of plants.

With annual, monotonous production of only wheat or corn coupled with intense tillage and summer fallowing, it is a heartbreaking tragedy not to realize that soil structure shall eventually be nearly or completely wiped out. But, it is also encouraging to realize that conservation tillage coupled with realistic extensions of no-tillage techniques during numerous years contributes to the conservation of soil structure, especially if crop rotation including pulse crops such as lentils and field pea along with small grain cereal crops is adopted. High harvests are accompanied by the increase of plant remnants, and thus a good base is laid for the increased production of glomalin and for increased aggregate stability.

Next, we mention steps aimed at the direct protection of soil against erosion. Crop residue cover should be kept until the crop canopy is closed, and thus the soil surface is protected against direct action of raindrops. Moreover, the presence of crop residues makes surface runoff more difficult owing to its increased flux resistance. Here, the word crop is meant in the broadest way as, e.g., the alternation of summer crops with winter crops and perennial crops.

Contour plowing, contour farming, and contour strip cropping are each very efficient to protect soil against erosion. The procedure described as contour plowing means that crops are planted nearly on contour lines. The ruts made by the plow run perpendicular rather than parallel to slopes, generally resulting in furrows that curve around the land and are level. These ridges along contour lines create a water-break that reduces the direct downhill slope flux and avoids the formation of rills as well as subsequent gullies during times of heavy water runoff. This practice is mostly effective on moderate slopes in ranges from 2 to 10 % when crops are planted in tilled soil with ridge heights of about 5–8 cm (2–3 in.). If the system is applied without plowing on moderate slopes, it is denoted as contour farming. It still reduces erosion, especially if plant residues cover the slope in horizontal strips.

Contour strip cropping involves strips of high-residue cover or perennial crops alternating with strips of low-residue cover. The strips having widths usually between 20 and 35 m (approximately 75 and 120 ft) should be positioned on the contour lines or close to them. Even though it is not always possible in rolling landscapes where the strips are interrupted by an abrupt obstacle, the technique is very useful. Soil that erodes from the bare or low-residue strips is deposited in strips with high residue or dense vegetation because runoff velocity is decreased. This practice is most useful if the soil is tilled or if the soil is left bare during part of the year in no-till. Soil eroding from bare or low-residue strips is deposited in strips with high-residue or dense vegetation because runoff velocity is decreased. The perennial contour strips build up a greater and more stable permeability thanks to the uninhibited activity of vegetation and fauna, so they are better at soaking up surface runoff and trapping sediment. This practice is most useful if the soil is tilled or if the soil is left bare during part of the year in no-till.

When both slope length and its steepness increase, runoff and soil loss also increase. Effective erosion protection can be best achieved if both characteristics are reduced. The slope steepness can be changed by the construction of terraces as is frequently done in southeastern Asia and around the Mediterranean. Because the terraces also shorten the length of slopes, they are a very effective contra-erosion measure but require a great deal of expense. They are constructed in such a way that crops can be grown on them. Terraces are designed to receive and store water until it infiltrates or is conducted by a canal to a safe outlet like a waterway. They are usually designed to drain completely in 48 h to avoid waterlogging within the terrace.

Generally, all measures and equipment leading to the reduction of erosion or its potential contribute to the healthy state of soils and to improved environmental conditions. The more we know about the soil and how it works, the better will be our designs and management.

Index

A
Acid soil reaction, 93
Acrisols, 200–202, 205, 206
Actinomycetes, 24, 25, 225
Active soil reaction, 94
Actual evaporation, 147, 148
Actual transpiration, 155, 156
Adenosine diphosphate (ADP), 161, 166
Adsorptive forces, 112
Aerobic bacteria, 24
Aerobic processes, 21, 46
Aggregate, 18, 27, 32, 61, 71–74, 76, 77, 85, 87, 88, 116, 125, 130, 132, 140, 144, 148, 150, 172, 194, 196, 199, 224, 227–230
A horizon, 72, 73, 80, 140, 173, 174, 181, 184, 187, 188, 190–192, 194, 195, 198–203, 205, 207–212, 216, 218, 219, 224, 225
Albaquals, 197
Albaquults, 197
Albeluvisols, 200
Albolls, 199
Alfisols, 197, 200–205, 207, 213
Algae, 27, 33, 164, 167, 168, 221
Alisols, 200, 201, 206, 211
Alkaliböden, 193, 194
Alkaline soil reaction, 93
Alluvial soils, 58, 177, 182, 185, 192
Al octahedron, 44
Amino group, 67, 76
Ammonification, 22, 165
Amphibole, 44, 203
Anaerobic bacteria, 24
Anaerobic processes, 21, 189
Andisols, 195, 205, 207

Andosols, 194, 195, 205, 207
Anthrosols, 189, 190
Ants, 9, 15, 29, 48, 64
Aquents, 192, 198
Aquepts, 192, 210
Aquoll, 192, 198
Arenic Rudosols, 203
Arenosols, 188, 202, 203, 205
Argiabolls, 197
Argissolos, 202
Aridisols, 193, 194, 199, 200, 205, 208, 213
Arizona, 9–11
ATP enzyme, 160, 161
Auenböden, 192
Autotrophic process, 159
Azonal soils, 204

B
Bacteria, 13, 22–27, 64, 99, 100, 164, 165, 168, 214, 221, 225
Bacteriophages, 22
B horizon, 73, 173, 174, 195, 196, 198–203, 205, 207, 211
Biodiversity, 15, 16, 18, 21, 30, 223
Biofuels, 16, 169, 221, 222
Biopores, 85
Bivalent ion, 24
Black Cotton Soils, 191
Black Dust Soils, 195
Blocky structure, 72
Borolls, 199
Braunerde, 201, 203
Brown Podzolic Soils, 202
Brown Soils, 199, 201, 203

© Springer Science+Business Media Dordrecht 2015
M. Kutílek, D.R. Nielsen, *Soil: The Skin of the Planet Earth*,
DOI 10.1007/978-94-017-9789-4

Printed in the United States
By Bookmasters